Environmental Subsidies to Consumers

A typical consumer underestimates the benefits of future energy savings and underinvests in energy efficiency, relative to a description of the socially optimal level of energy efficiency. Various programs have been implemented to alleviate this energy efficiency gap problem. In recent years, many governments have started providing consumers with subsidies on the purchases of eco-friendly products such as hybrid cars and energy-efficient appliances. This book conducts a comprehensive analysis of the environmental subsidy programs conducted in Japan and examines their impacts on consumer product selection, consumer product use, and environmental outcomes. The book also proposes recommendations for future environmental and industrial policies.

The book's empirical findings will be of interest to those who are performing research in this area and those who are involved in developing environmental and industrial policies.

Shigeru Matsumoto is Professor at the Department of Economics, Aoyama Gakuin University. He studied on a Heiwa Nakajima Foundation Scholarship at North Carolina State University, where he earned his PhD in Economics. His research interest lies in applied welfare economics, with particular focus on consumer behavior analysis.

W0234935

Environmental Subsidies to Consumers

How did they work in the
Japanese market?

Edited by Shigeru Matsumoto

Routledge
Taylor & Francis Group

LONDON AND NEW YORK

First published 2016
by Routledge
2 Park Square, Milton Park, Abingdon, Oxfordshire OX14 4RN

and by Routledge
711 Third Avenue, New York, NY 10017

First issued in paperback 2017

Routledge is an imprint of the Taylor & Francis Group, an informa business

British Library Cataloguing in Publication Data
A catalogue record for this book is available from the British Library

Library of Congress Cataloging-in-Publication Data
Environmental subsidies to consumers : how did they work in the
    Japanese market? / edited by Shigeru Matsumoto.
        pages cm
    1. Environmental policy—Economic aspects—Japan.   2. Subsidies—
Japan.   3. Energy tax credits—Japan.   4. Energy consumption—
Economic aspects—Japan.   I. Matsumoto, Shigeru.
    HC465.E5E585 2015
    338.952′02—dc23
    2015000462

ISBN 13: 978-1-138-06731-8 (pbk)
ISBN 13: 978-0-415-73107-2 (hbk)

Typeset in Galliard
by Apex CoVantage, LLC

Contents

Figures

Part IV

Tables

Part IV

Contributors

Toshi H. Arimura is Professor at the Department of Political Science and Economics of Waseda University.

Keiko Hirota is Global Networking and Public Relations of Japan Automobile Research Institute.

Kazuyuki Iwata is an Associate Professor at the Department of Regional Policy of Takasaki City University of Economics.

Shigeru Kashima is Professor at the Graduate School of Science and Engineering of Chuo University.

Shigeru Matsumoto is Professor at the Department of Economics, Aoyama Gakuin University.

Kenichi Mizobuchi is an Associate Professor at the Department of Economics of Matsuyama University.

Minoru Morita is a Junior Researcher at the Research Institute for the Environment and Trade, Waseda University.

Satoshi Sekiguchi is a Professor at the Department of Economics of Rikkyo University.

Kenji Takeuchi is a Professor at the Graduate School of Economics of Kobe University.

Tomohiro Tasaki is Head of the Center for Material Cycles and Waste Management Research at the National Institute for Environmental Studies.

Keiko Yamaguchi is an Associate Professor at the Graduate School of Humanities and Social Sciences of Okayama University.

Acknowledgments

This work was supported by the Kondo Jiro Grant of Asahi Glass Foundation and a Grant for Environmental Research Projects from the Sumitomo Foundation. The authors highly acknowledge their financial supports.

Acronyms and abbreviations

3Rs	reduce, reuse, and recycle
AEHA	Association for Electric Home Appliances
AMTD	average monthly travel distance
CCA	command and control approach
CRT	cathode ray tube
ECCJ	Energy Conservation Center Japan
EDMC	Energy Data and Modeling Center
DSM	demand-side management
EAP	eco-action point
EIA	economic incentive approach
EPP	eco-point program for appliances
EU	European Union
FIT	feed in tariff
HV	hybrid vehicle
IEL	iso-environmental-load
IRS	Internal Revenue Service
ISO	International Organization for Standardization
JAMA	Japan Automobile Manufacturers Association
LCA	life-cycle assessment
METI	Ministry of Economy, Trade and Industry
MIAC	Ministry of Internal Affairs and Communications
MLIT	Ministry of Land, Infrastructure, Transport and Tourism
MOE	Ministry of the Environment
MPMHPT	Ministry of Public Management, Home Affairs, Posts, and Telecommunications MPMHPT
OECD	Organization for Economic Co-operation and Development
PPGS	photovoltaic power-generation system
PPP	polluter pays principle
U.S.	United States of America
U.K.	United Kingdom
WTO	World Trade Organization

Part I
Environmental subsidies to consumers: conceptual issues

1 Environmental subsidies to consumers

Introduction

Shigeru Matsumoto

1.1. Objective of the book

Households consume a significant amount of energy and thus have a substantial impact on the local as well as global environment. Including electricity consumption as their primary source of energy, U.S. households account for 21 percent of the country's total energy consumption (U.S. Energy Information Administration 2015). In their review article, Tukker and Jansen (2006) show that the final energy consumption and environmental burden in European countries is mostly driven by the transportation and household sectors. Furthermore, the OECD (Organization for Economic Co-operation and Development 2008) projects that the impact of the household sector on natural resource stocks, environmental quality, and climate change will increase by 2030. The impact of the household sector in non-OECD countries is expected to be considerably greater because of their high economic growth. This scenario suggests that sustainable consumption is a major challenge for our society.

Thus, to achieve sustainable consumption, we must improve energy efficiency at the household level. Manufacturers have invested intensively in the research and development of energy efficient household products for the past several decades, and thus the energy efficiencies of such products have improved considerably. For example, from 1990 to 2013, the energy consumption of new upright freezers sold in the United Kingdom fell by 60 percent and that of new refrigerators by 57 percent (Department of Energy and Climate Change of U.K. 2014). Despite such significant improvement in the energy efficiency of home electric appliances, the total energy consumption of households has not decreased correspondingly. Although some home appliances have become more energy efficient, the increased number of such devices has offset these efficiency gains (U.S. Energy Information Administration 2013).

Similarly, in the transportation sector in Japan, the average fuel economy of gasoline vehicles sold improved by 71.5 percent from 1995 to 2012 (Ministry of Land, Infrastructure, Transport and Tourism [MLIT] of Japan 2013). Despite this improvement, total gasoline consumption increased by 9.3 percent during this period (Agency for Natural Resources and Energy of Japan 2014).

These previous experiences suggest that an improvement in the energy efficiency of household products is necessary but not sufficient for reduced energy consumption. A better understanding of household behavior is also required to achieve sustainable consumption.

Households seldom recognize the quantum of electricity they use for home appliances. If electricity consumption information is displayed while appliances are in use, households may alter their behavior. Smart meters enable households to access accurate and detailed information about their electricity use. Oxford Economics (2012) estimates that in the United Kingdom, using smart meters would result in savings of £8.978 billion ($14.156 billion) from 2012 to 2030,[1] which means a yearly savings of 5 percent for an average household. For this reason, the U.K. government intends to equip all homes and small businesses with smart meters by 2020 (Gov.UK 2014).

The energy efficiency of home products has to be presented in a way that households can easily understand. To this end, ENERGY STAR labels developed by the U.S. Environmental Protection Agency in 1992 are now used internationally. By counting the number of stars on the label, households can easily understand the annual electricity consumption of an appliance. Similar energy efficiency labels are now used for other home products. Homebuyers (or renters) will not know the energy efficiency of a building at the time of purchase (or lease signing). To inform them about this fact, energy performance certificates were introduced in the European Union (EU) in 2002. The certificate displays the energy performance of a building with a comparison label, similar to the EU Energy Star Label (Amecke 2011).

The provision of energy efficiency information might alter the energy consumption behavior of households. However, such change will not occur in a perfectly rational manner. Since a typical household underestimates the benefit of future energy savings, it underinvests in energy efficiency relative to the socially optimal level. To alleviate this energy efficiency gap problem, a wide variety of approaches have been undertaken in recent years, including the introduction of environmental subsidies to consumers.

Many governments subsidize the purchase of eco-friendly vehicles. The U.S. Energy Policy Act of 2005 provides a credit for taxpayers who purchase certain energy efficient vehicles, including qualified hybrid vehicles (U.S. Internal Revenue Service [IRS] 2007). In 2014, plug-in and electric vehicles were eligible for a maximum federal tax credit of $7,500 (U.S. Department of Energy 2014b). European countries have implemented incentive programs promoting the use of eco-friendly vehicles and the scrapping of obsolete, polluting vehicles. France uses a bonus–malus program in which the maximum benefit paid to new e-car buyers is $7,000, while the buyers of high-carbon-emitting vehicles incur penalties of up to $3,650 (Hockenos 2011). Similarly, in 2006, the U.K. tax rate for the least polluting cars (brand A: 0–100 g/km) was reduced to zero, while the tax rate for the most polluting cars (brand G: 226 g/km and higher) was increased to £300 ($474) in 2007 and to £400 ($632) in 2008 (OECD 2011).

The U.S. Department of Energy (2014a) announced in July 2009 that $300 million in funding from the American Recovery and Reinvestment Act would be made available to states and territories to promote the purchase of ENERGY

STAR–qualified appliances. The appliances eligible for rebates include central air conditioners, heat pumps, boilers, furnaces, room air conditioners, clothes washers, dishwashers, freezers, refrigerators, and water heaters (Energy Star 2014). The Chinese government offered RMB 26.5 billion ($4.17 billion) to the subsidy program for energy efficient appliances.[2] The buyers of energy-saving water heaters, refrigerators, and washing machines received as much as RMB 400 ($63) (Bloomberg News 2012).

A majority of the energy used in homes is for space heating and cooling. For instance, 48 percent of the energy consumption of U.S. homes in 2009 was used for space heating and cooling (U.S. Energy Information Administration 2013). If the energy efficiency of buildings is improved, household energy consumption will decrease substantially. In Canada, the ecoENERGY Retrofit program provided grants of up to $5,000 to help homeowners increase energy efficiency at home and reduce the burden of high energy costs (Canada.ca 2014). Under the Warm Front Programme, the United Kingdom earmarked £1.2 billion for improving energy efficiency in poor households (OECD 2008).

The above examples show that subsidy programs are becoming increasingly popular. Despite this popularity, their economic and environmental impacts have not been examined in detail. This volume reviews Japan's environmental subsidy programs and examines their impact on consumer product selection and usage and environmental outcomes. Although we focus on Japan, we believe that our empirical findings are useful for international readers as well. As we mentioned, similar subsidy programs have been used in many countries.

1.2. The structure of the book

The book is divided into five parts. Part I, which includes this introduction, deals with conceptual issues. After briefly reviewing the environmental subsidy programs implemented in major countries, we summarize the debate on these programs in Chapter 2. In recent years, environmental subsidies to consumers have become increasingly popular in Japan. Chapter 3 explains the factors behind this trend.

Part II analyzes the impacts of various rebate programs for energy efficient appliances. The Japanese government introduced a cash rebate program called the "eco-point program for appliances." In this program, consumers who purchased energy efficient digital televisions, washing machines, and air conditioners obtained "eco points" that could be used to buy other goods and services. By providing eco points only for energy efficient appliances, the Japanese government attempted to reduce electricity consumption from the use of appliances.

Chapter 4 examines the effect of the eco-point program on consumers' TV selection. We find that the eco-point program induced consumers to choose not only energy efficient TVs but also large screen TVs. In summary, the eco-point program increased the total electricity consumption from TV usage. Since appliances are replaced after several years of use, they have a replacement cycle that may be affected by such programs. Chapter 5 considers the replacement cycle of refrigerators and evaluates their CO_2 reduction resulting from the eco-point program.

Knowledge about household appliance usage is essential for the evaluation of actual energy efficiency. Chapter 6 describes a household survey examining whether households change their space-cooling behavior after purchasing a new energy efficient air conditioner. To describe an energy efficiency gap problem, previous studies have estimated an "implicit discount rate" that consumers reveal through their product purchases has been estimated. Chapter 7 evaluates the effect of the eco-point program for appliances on the implicit discount rate.

Part III focuses on rebate programs for eco-friendly vehicles. A wide variety of eco-car policies can be found in Asian countries. Chapter 8 compares those eco-car policies and discusses their weaknesses and strengths. The Japanese government lowered vehicle-related taxes and offered a cash rebate for eco-friendly vehicles. Chapter 9 analyzes the "eco-car subsidy program" recently implemented in Japan and estimates the impact of the program on vehicle usage.

Part IV discusses environmental subsidy programs from a broader perspective. Chapter 10 studies the eco-point program for energy efficient houses. This program will alter the quality of housing stock and may substantially improve energy efficiency in the housing sector. The chapter shows that the program will not be a cost-effective policy to address global warming. Chapter 11 reassesses environmental subsidy programs from a life cycle perspective. Although energy consumption measured by product usage is used as an efficiency criterion in most such programs, it is necessary to consider the environmental burden generated throughout the entire life of a product. Chapter 11 offers an alternative efficiency criterion assessed from a life cycle perspective.

Part V summarizes the findings in this volume and suggests environmental and industrial policies for the future.

Notes

1 £1 is converted into $1.58.
2 RMB 1 is converted into $0.16.

References

Agency for Natural Resources and Energy of Japan. (2014) Energy Consumption Statistics. Available: <http://www.enecho.meti.go.jp/statistics/energy_consumption/ec001/> (accessed 12 November 2014).

H. Amecke. (2011) 'The Effectiveness of Energy Performance Certificates – Evidence from Germany', CPI report. Climate Policy Initiative.

Bloomberg News. (5 June 2012) 'China Subsidizes Purchases of Energy-Saving Appliances', Bloomberg News.

Canada.ca. (2014) 'ecoENERGY Retrofit – Homes Program', Available: <http://www.nrcan.gc.ca/energy/efficiency/housing/home-improvements/5003> (accessed 15 November 2014).

Department of Energy and Climate Change of the United Kingdom. (2014) 'Domestic energy consumption in the U.K. between 1970 and 2013', Chapter 3, *Energy*

Consumption in the UK 2014, Available: <https://www.gov.uk/government/uploads/system/uploads/attachment_data/file/338662/ecuk_chapter_3_domestic_factsheet.pdf> (accessed 12 November 2014).

Energy Star. (2014) 'Special Offers and Rebates from ENERGY STAR Partners', Available: <http://www.energystar.gov/rebate-finder> (accessed 12 November 2014).

Gov.UK. (2014) 'Helping Households to Cut Their Energy Bills', Available: <https://www.gov.uk/government/policies/helping-households-to-cut-their-energy-bills/supporting-pages/smart-meters> (accessed 4 November 2014).

P. Hockenos. (2011) 'Europe's Incentive Plans for Spurring E.V. Sales', *New York Times,* Available: <http://www.nytimes.com/2011/07/31/automobiles/europes-incentive-plans-for-spurring-ev-sales.html?_r=1&> (accessed 18 November 2014).

MLIT of Japan. (2013) 'Change in the average fuel economy of gasoline vehicles', Available: <http://www.mlit.go.jp/common/000990330.pdf> (accessed 7 November 2014).

OECD. (2008) *Environmental Outlook to 2030.* OECD, Paris.

OECD. (2011) *Invention and Transfer of Environmental Technologies.* OECD Studies in Environmental Innovation. OECD, Paris.

Oxford Economics. (2012) The Value of Smart Metering to Great Britain, Available: <http://www.britishgas.co.uk/content/dam/british-gas/documents/Oxford-Economics-Infographic_final.pdf> (accessed 10 November 2014).

A. Tukker, and B. Jansen. (2006) 'Environmental impacts of products: A detailed review of studies', *Journal of Industrial Ecology* 10, 159–82.

U.S. Department of Energy. (2014a) 'State Energy-Efficient Appliance Rebate Program', Available: <http://www1.eere.energy.gov/recovery/appliance_rebate_program.html> (accessed 12 November 2014).

U.S. Department of Energy. (2014b) 'Tax Incentives Information Center', Available: <http://www.fueleconomy.gov/feg/taxcenter.shtml> (accessed 10 November 2014).

U.S. Energy Information Administration. (2015) 'Drivers of U.S. Household Energy Consumption, 1980–2009', Available: <http://www.eia.gov/analysis/studies/buildings/households/> (accessed 25 March 2015).

U.S. IRS. (2007) 'Summary of the Credit for Qualified Hybrid Vehicles', Available: <http://www.irs.gov/uac/Summary-of-the-Credit-for-Qualified-Hybrid-Vehicles> (accessed 10 November 2014).

2 Environmental subsidies to consumers as policy instruments

Shigeru Matsumoto

2.1. Introduction

Many countries have started implementing various environmental subsidy programs. Consumers can now receive subsidies when purchasing eco-friendly houses, vehicles, and appliances. Why have such environmental subsidy programs to consumers become popular in recent years?

The environmental burden of household consumption is expected to continue to grow. For instance, OECD (2011) estimates that the energy consumption in the household sector of OECD countries will increase at an annual rate of 1.4 percent between 2003 and 2030, while waste generation will increase at an annual rate of 1.3 percent. The rates among non-OECD countries are expected to become much higher. Therefore, it is very important to provide households with an incentive to reduce their environmental burden.

Economists often recommend using environmental taxes to promote environmentally friendly consumption. However, environmental taxes rarely obtain public support, since the taxes tighten the household's budget constraint. On the other hand, environmental subsidies are more likely to be supported by public opinion, since the subsidies relax the budget constraint. If politicians state that a planned subsidy program will revitalize the country's economy, then the program will obtain public support more easily.

Although environmental subsidies to consumers have become rapidly popular in recent years, their effectiveness has not been carefully examined. In this chapter, we provide a brief overview of the major environmental subsidy programs recently introduced in Japan and discuss their effectiveness.

The structure of the chapter is as follows. The traditional opinion about environmental subsidies is negative – that is, environmental subsidies should not be used for pollution management. We explain such a traditional opinion in Section 2.2. Despite the opposition, environmental subsidies have been widely used in the real world. We explain the logic behind environmental subsidies in Section 2.3. A variety of approaches are used in environmental subsidy programs. In Section 2.4, we classify environmental subsidy programs according to their approach and explain their differences. In Section 2.5, we briefly describe the major environmental subsidy programs recently introduced in Japan and evaluate their effectiveness. In Section 2.6, we outline the future research agenda for environmental subsidy programs and conclude the chapter.

2.2. Traditional opinion on environmental subsidies

2.2.1. *Efficiency of the economic incentive approach*

Traditionally, command and control approaches (CCAs) have been used for environmental regulation. CCAs solve environmental problems by restricting activities that are directly harmful to the environment. Although environmental economists have pointed out the potential problems of CCAs, these approaches remain the most popular form of environmental regulation.[1]

Under a CCA, all polluters are required to take the same environmental measure regardless of the implementation cost. Although such a uniform measure simplifies enforcement of the regulation, the total abatement cost that a society spends to reduce a given amount of pollutants becomes more expensive. This is because CCAs do not allow regulatory adjustment, which softens the regulatory standard for polluters that have high abatement costs and tightens the regulatory standard for polluters that have low abatement costs.

CCAs are often used for the management of toxic substances that can damage human health, since the reliability of pollution control is prioritized. In contrast, for a pollutant that does not cause severe environmental damage immediately, its efficient management and control can be more open to discussion. In such a case, the government intends for the total amount of pollutants released into the environment to not exceed a certain threshold level. For the management of such non-dangerous pollutants, economic incentive approaches (EIAs) can be adopted.

The two most popular EIAs are environmental taxes and transferable permits. It is well known that both approaches lead to the same environmental outcome (Hosoda 2000). Under environmental taxes, each polluter compares the tax for pollution emission and the abatement cost. Then the polluter adjusts the emission level so that the marginal abatement cost equals the tax rate. A policymaker transforms the unit price of pollution emission into a tax rate and equalizes the marginal abatement cost among polluters indirectly. Consequently, an environmental tax EIA can minimize the total abatement cost to society (Baumol and Oates 1971).

In reality, a policymaker considers not only environmental policies, but also industrial, employment, and regional policies. The implementation of an environmental tax will not affect all polluters equally. By examining the regulatory standard across industries, Matsumoto and Yokoyama (2008) found that the policymaker intentionally differentiates the emission standard of SO_2 across industries.

Once they have implemented a measure required under a CCA, polluters will not take additional actions to reduce their pollution emission because they will derive no benefit from doing so. In contrast, under EIAs, polluters can continue to reduce their tax payment if they further reduce their pollution emission. Therefore, they keep engaging in pollution abatement. Consequently, even if the reduction of pollution emission through EIAs is small in the short run, it can be large in the long run since it will change both household and firm behavior (Stavins 2003). Anderson and Newell (2004), Hassett and Metcalf (1995), and Jaffe et al. (1995) report that the increase in energy prices has promoted investment in energy-saving initiatives, while Popp (2002) reports that the increase has induced technological innovation. These studies provide empirical evidence that support the long-run effectiveness of EIAs.

2.2.2. Conventional economic approaches vs. subsidies

How do environmental subsidies differ from other economic approaches such as environmental taxes and emissions trading? In this section, we examine their differences.

In 1972, the OECD Council adopted the polluter pays principle (PPP) as a general principle of environmental policy (OECD 1972). This principle states that the cost of environmental damage needs to be borne by the polluters. OECD (1974) further states that OECD member countries should not assist the polluters in bearing the costs of pollution control, whether by means of subsidies, tax advantages, or other measures.

Following this principle, countries' official statements are that they have not provided subsidies to polluters. For instance, in a 1980 environmental white paper, the Japanese government stated that it had not adopted any subsidy initiatives to resolve pollution problems.

> After various pollution regulations primarily focusing on emissions control were introduced in 1970, private companies rapidly increased investment in pollution control. However, the government has provided neither capital investment costs nor recurrent costs for pollution prevention.
> (Ministry of the Environment [MOE] 1980, Chapter 3)

There are two major reasons why subsidies should not be used for pollution management. The first reason is that environmental subsidies distort resource allocation across industries. If polluters can financially benefit by taking a pollution measure, they will take it. A policymaker can provide polluters with a pollution abatement incentive with environmental subsidies. Under certain conditions, a policymaker can make each polluter take the same pollution measure as it would under an environmental tax. Nevertheless, an environmental subsidy leads to a different long-run equilibrium than the one under an environmental tax. An environmental tax increases the operating cost of polluting industries by taking out an economic rent. In contrast, a subsidy offers an economic rent to polluting industries and lowers their operating cost. The differences in operating costs affect the entry–exit condition of polluting industries, and thus the expected long-run equilibrium differs between the environmental tax and subsidy scenarios. In fact, it is possible to find a situation in which an environmental subsidy invites more firms into the polluting industry and actually worsens the pollution problem (Baumol and Oates 1988).

When the Japanese government introduced an environmental tax in October 2011, there was a debate on how to spend the tax revenue. Some argued that the tax revenue collected should be spent on global warming prevention measures in the industry bearing the environmental tax. Although such a request sounds reasonable, it ignores an aspect of the long-run equilibrium. One of the main purposes of environmental taxation is to remove the economic rent of polluting industries. Therefore, the collected tax revenues are not required to be spent on global warming measures.

The World Trade Organization (WTO) specifies the rules of application to subsidy programs in the Agreement on Subsidies and Countervailing Measures (WTO 2014), since subsidies can distort free trade by protecting domestic industries. The second reason for traditional opposition to environmental subsidies is that they can distort international trade and investment. It is easy to find a claim that environmental subsidies have distorted international trade in the real world – for instance, the U.S. Department of Commerce (2014) announces that it will impose anti-subsidy duties against U.S. imports of Chinese solar technology products.

Nonetheless, not all environmental subsidies are prohibited. In fact, OECD (1974) states that an environmental subsidy is acceptable if it complies with all of the following conditions:

a)   It should be selective and restricted to those parts of the economy, such as industries, areas, or plants, where severe difficulties would otherwise occur;

b)   It should be limited to well-defined transitional periods, laid down in advance, and adapted to the specific socio-economic problems associated with the implementation of a country's environmental programme;

c)   It should not create significant distortions in international trade and investment.

(OECD 1974, Guiding Principles)

Because of the existence of the exception rule, countries have adopted various environmental subsidies. For instance, in a 1998 environmental white paper, the Japanese government stated that it would promote subsidy programs for pollution mitigation:

To further mitigate pollution problems by installing abatement technologies in firms, the government considers the characteristics of the Japan Environment Corporation, the Japan Development Bank, and the Japan Finance Corporation for Small Business and then promotes their subsidy programs.

(MOE 1998, Chapter 4)

The PPP defines polluters as not only producers but also consumers. In recent years, cases where environmental subsidies to consumers have distorted free trade have been pointed out. For instance, Kawase (2011) examines the case of the ecocar, an initiative recently introduced by the Japanese government to promote ecocars. Kawase argues that the host country has used the eco-car subsidy program to discriminate against foreign automobile makers when considered in light of the manner of the subsidy's implementation.

What conditions have made environmental subsidies to consumers popular in recent years? In the next section, we will review these conditions.

2.3. Why are environmental subsidies to consumers used?

Economists explain that an externality, one type of market failure, causes environmental problems and then recommend using EIAs to manage pollution. However, there are situations in which other market failures exist or the assumptions of the economic analysis are not satisfied within specific polluting industries. The resolution of other market failures is often mentioned as a reason for environmental subsidies to consumers. We will explain below.

2.3.1. Support of new technology

When a new eco-friendly product is released to the market, it is sold at a high price. Only consumers with strong concern for the environment purchase it. After a producer succeeds in production innovation and lowers the price, ordinary consumers consider a purchase of the product. At this stage, they can consult their friends who have already purchased the product to learn more about it. Once the price of the product falls below a certain threshold, the eco-friendly product becomes a competitive product. Subsequently, the market share of the product begins to increase.

Some eco-friendly products not only reduce the burden on the environment but also provide a financial benefit to consumers. Those include durable goods that require energy for use. Typical examples are home electric appliances, such as air conditioners and refrigerators. Energy-saving products are not only friendly to the environment but also friendly to the consumer's wallet. The product penetration process mentioned in the previous paragraph applies to such energy-saving products.[2]

Although Fernando Porsche developed a hybrid vehicle technology more than 100 years ago, it has not been used until recently. Toyota succeeded in its mass production of the Prius in 1997, while Honda succeeded in that of the Insight in 1999. However, the market share of hybrid vehicles remained low until the mid-2000s. After the price of hybrid vehicles dropped to 3 million yen ($25,000), they became competitive with conventional gasoline vehicles.[3] With the introduction of the eco-car subsidy program, they rapidly penetrated into the market.

The economy of scale of eco-friendly products and the promotion of the new environmental technology are often used to justify environmental subsidy programs.

2.3.2. Relaxing a liquidity constraint

A person lacking wealth cannot make a loan arrangement to purchase durable products. He or she has to choose an affordable product given the cash on hand. The liquidity constraint prohibits the person from choosing an energy-efficient product even if it would be beneficial for him or her in the long run.

Firms can borrow money on the security of future energy savings. However, households cannot make such an arrangement when purchasing a durable good because the intensity of the good's use differs widely among households.

In the presence of liquidity constraints, the demand for eco-friendly products becomes lower than the socially desirable level. Environmental subsidy programs are expected to relax the liquidity constraint.

2.3.3. Provision of product information

When purchasing a product, a consumer will not know all of the product's characteristics. For instance, the annual electricity cost for electric appliances can be found in the store in developed countries. However, such information is not available in many developing countries. Therefore, a consumer in a developing country does not have access to information regarding appliances' full costs and chooses energy-intensive products sold at lower prices.

Once consumers start using the product, they will know its energy efficiency. However, they still cannot know all the product's characteristics even after they use it. Some products are produced through an environmentally friendly production process, while other products are produced through a pollution-intensive production process. Consumers cannot identify the production process, even if they use the product. The production process and other product characteristics that consumers cannot know after consumption are called credence attributes (Caswell and Mojduszka 1996). Credence attributes are important in the purchase of eco-friendly products.

A policymaker can provide product information to encourage consumers to purchase eco-friendly products. However, a typical consumer spends only a limited amount of time shopping and does not carefully examine all product information. Therefore, a policymaker must carefully design the information and the manner in which it is provided. Many ideas, such as eco-labeling, have been developed for this purpose.

2.3.4. Remedy for bounded rationality

If adequate information is provided, will consumers choose a product rationally? Most economists have provided a negative answer to this question.

For example, when a consumer purchases an electric appliance, he or she compares the price of the appliance with the subsequent expected electricity cost. Therefore, by comparing the sales price of the appliance with the annual electricity cost, we can show how an average consumer evaluates future energy savings.

Suppose that there are two electric appliances a and b whose sales prices are P_a and P_b. For the sake of simplicity, we assume that both appliances provide the same service. The annual electricity cost of appliance a is E_a, while that of appliance b is E_b. It is assumed that $E_b > E_a$ – that is, appliance a is more energy efficient than appliance b. Now assume that the period of use is T years. Then we can calculate the net present value of future energy savings as

$$P_a - P_b = \sum_{t=0}^{T} \frac{E_b - E_a}{(1 + \theta/100)^t}.$$

The discount rate θ that satisfies the above equality is called the implicit discount rate. This discount rate is often used for consumers' evaluation of energy investment.

A high implicit discount rate implies that consumers put more weight on the upfront purchase cost than on the benefit from future energy savings. Past empirical studies have estimated that implicit discount rates for electric appliances can reach from 25 to 100 percent (Sanstad et al. 2006). These rates are much higher than the discount rate generally applied in economic analyses.[4] A consumer having a high implicit discount rate underinvests in energy savings. This phenomenon is called the energy efficiency gap.

The energy efficiency gap indicates that consumers do not make a rational choice from a long-run point of view, even if adequate information is available. Environmental subsidies are used to correct this myopic consumer behavior.

2.4. Types of environmental subsidies

When market or behavior failures exist, a policymaker can enhance social welfare by introducing subsidy programs. Although a wide variety of subsidy programs have been used for the past several decades in many countries, they can be broadly classified as follows.[5]

2.4.1. Discount sales

A policymaker provides subsidies directly to consumers so that they can purchase eco-friendly products at discounted prices. Since a policymaker cannot directly control the sales price, the subsidy works by providing a rebate to a consumer who fills out an application form when he or she buys the product. Thus, the method of discount sales is like a cash-back campaign conducted by the

government, through which the benefit of eco-friendly products is easily identified by consumers.

2.4.2. Preferential interest rate

Consumers sometimes borrow money from public or semi-public institutions
to purchase a durable good. Many institutions offer a preferential interest rate
when purchasing eco-friendly products. By offering a lower interest rate, a
policymaker induces consumers to purchase eco-friendly products. For instance,
a household makes a loan arrangement at the Housing Loan Corporation to
buy a new house. The corporation offers a preferential interest rate to the
household if it builds an eco-friendly house. Unlike the discount sale previously
mentioned, consumers receive benefits for several years.

Tax reductions

A policymaker allows consumers to pay reduced taxes when they purchase eco-
friendly products. Sales or consumption taxes are reduced in some cases, while
income tax is reduced others. To provide consumers with a sufficient incentive,
policymakers generally allow tax reductions for expensive durable products such
as vehicles and houses.

2.4.3. Comparison of subsidy schemes

When a policymaker uses either price reduction or preferential interest rate, he
must take certain political actions to manage the budget for the program. This
requirement causes practical problems. Although a subsidy program concludes
when the budget is exhausted, households cannot precisely forecast the end
date. To avoid confusion, a policymaker needs to report the balance of the
subsidy on a regular basis. Furthermore, there is a replacement cycle for a
durable good. If a subsidy program lasts into the replacement period, then a
consumer uses the program. Otherwise, he or she will not use it.

The environmental efficiency of subsidy programs has been analyzed in the
academic literature. In particular, demand-side management (DSM) of electricity
consumption has been extensively analyzed (Gillingham et al. 2009). However,
the evaluation of programs varies across studies. Some scholars, such as Loughran
and Kulick (2004), report that DSM was inefficient at improving energy conservation, while other scholars, such as Auffhammer et al. (2008) and Arimura
et al. (2012), report that DSM improved it efficiently.

A small number of studies have compared the efficiency of subsidy programs.
Gallagher and Muehlegger (2011) examined the effectiveness of the subsidy
program for the promotion of hybrid vehicles, and found that the reduction of
the sales tax had been more effective than deducting income tax after the purchase. Perhaps consumers are very concerned about upfront cost, and thus

subsidy programs that reduce upfront costs provide a stronger incentive to purchase eco-friendly products.

2.5. Environmental subsidies to Japanese consumers

In this section, we review environmental subsidy programs recently introduced in Japan and evaluate the effectiveness and problems of the programs.

2.5.1. Eco-car subsidy program

Two series of eco-car subsidy programs have been implemented to reduce carbon emissions associated with vehicle use. At the same time, the programs also aim to revitalize the country's economy. During the program periods, the buyers of ecocars received subsidies. The period of the first program was June 2009– September 2010, while the period of the second program was April 2012– September 2012.

The first program was initially planned between June 2009 and March 2010, with a supplementary budget of 357.2 billion yen ($2.98 billion). However, an additional supplementary budget was compiled for 230.4 billion yen ($1.92 billion), and the period of the program was extended until September 2010. After going through a required process, buyers receive a subsidy of 100,000 yen ($833.33) for a regular vehicle (70,000 yen or $583.33 for a light vehicle). If they replaced an old vehicle (whose age exceeded 13 years) with an eco-friendly vehicle, then they received 250,000 yen ($2,083.33). In the second installment of the program, the scrap incentive program was abolished and only the eco-car subsidy was provided. The total budget of the second program was 300 billion yen ($2.5 billion) – about half of the first program's budget. The number of the applications in the first program was 5.51 million, while those in the second program numbered 2.96 million ($246,700) (Ministry of Economy, Trade and Industry [METI] 2012).

In addition to the provision of subsidies, the Japanese government reduced vehicle-related taxes on ecocars. Consumers have to pay an acquisition tax corresponding to 5 percent of the acquisition price of a vehicle; but when the program was implemented, the acquisition tax was exempted for the purchase of hybrid vehicles. Vehicle owners are required to pay a weight tax to the country and an automobile tax to the prefecture. These taxes were also reduced for ecocars. The benefit of tax reduction was larger when purchasing a large, expensive ecocar.

Figure 2.1 shows the historical change in domestic sales of new vehicles, based on the data provided by the Automobile Information Center (2013). The year-to-year rate is shown on the vertical axis. The two arrows in the figure show the periods of the eco-car subsidy program. The figure demonstrates that the sales of new vehicles rapidly increased during the first program period (from 2009 to 2010) and sharply decreased after the program ended. The figure also shows that the sales of new vehicles rapidly increased again after the second

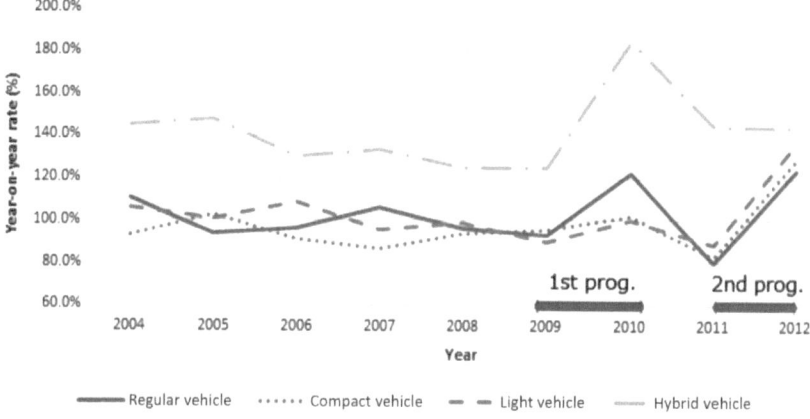

Figure 2.1 Historical change in domestic sales of new vehicles
Source: Automobile Information Center (2013)

subsidy program. We can also confirm that the sales of hybrid vehicles almost doubled during the first program. Therefore, eco-car subsidy programs stimulate the country's economy not only by stimulating the sales of vehicles but also by contributing to the market penetration of hybrid vehicles. The program therefore achieves its objective.

2.5.2. Eco-point program for appliances

The Japanese government implemented the program to promote the spread of green home appliances by utilizing eco points, the so-called "eco-point program for appliances," from 15 May 2009 to 31 March 2011. The program aimed to push ahead with anti–global warming projects, revitalize the economy, and encourage the diffusion of digital terrestrial televisions.

In this program, the buyers of green home electronics (energy-saving terrestrial digital broadcasting-compatible TVs, air conditioners, and refrigerators) obtained eco points that were exchangeable for a wide variety of products. When replacing old appliances, the buyers obtained recycle points until 31 December 2010. Eco points were reduced to half after 1 December 2010, and the energy efficiency criteria were tightened after 1 January 2011.

There were 45.49 million applications for the eco-point program, of which 32.30 million were for digital televisions, 7.29 million were for air conditioners, and 5.21 million were for refrigerators. The total number of eco points issued was worth 639.5 billion yen ($5.33 billion), of which the points for digital televisions equaled 52.20 billion yen ($0.44 billion), those for air conditioners equaled 61.6 billion ($0.51 billion), and those for refrigerators equaled 55.9 billion yen ($0.47 billion). One eco point is worth one yen. According to the MOE (2012), 97.93 percent of eco points were exchanged with gift certificates or prepaid cards.

Consumers who purchased a large appliance received more points; that is, they received more points when purchasing a large-screen digital television, an air conditioner designed for a large room, or a large-capacity refrigerator. If a consumer purchased an air conditioner for the living room over 33 m² or a refrigerator designed for a household of 3–4 persons (i.e., 400 liters or larger), then he or she received 9,000 yen ($75). If he purchased a digital television with a 40-inch screen, then he received 23,000 yen ($191.66). The buyer of a digital television received the most eco points.

Figure 2.2 shows the change in the sales of air conditioners, refrigerators, and digital televisions (METI 2013). It demonstrates that the yearly sales of digital televisions exceeded 100 percent in 2008. Therefore, an upward trend of digital television sales had been already observed before the eco-point program. Households needed to prepare for the transition from the analog TV broadcast to the digital broadcast. In contrast, the yearly sales of air conditioners and refrigerators in 2009 were below 100 percent. The sales of the three appliances rapidly increased after the summer of 2010 and reached a peak in December of 2010, that is, just before the reduction in the value of eco points. After that, the sales of the three appliances dropped sharply.

The figure also shows that the impact of the eco-point program on sales is large in the order of television, air conditioner, and refrigerator. At the peak, sales of televisions, air conditioners, and refrigerators achieved six times, three times, and two times the value of sales before the eco-point program, respectively. Hence, we can state that the impact of the eco-point program on sales was substantial.

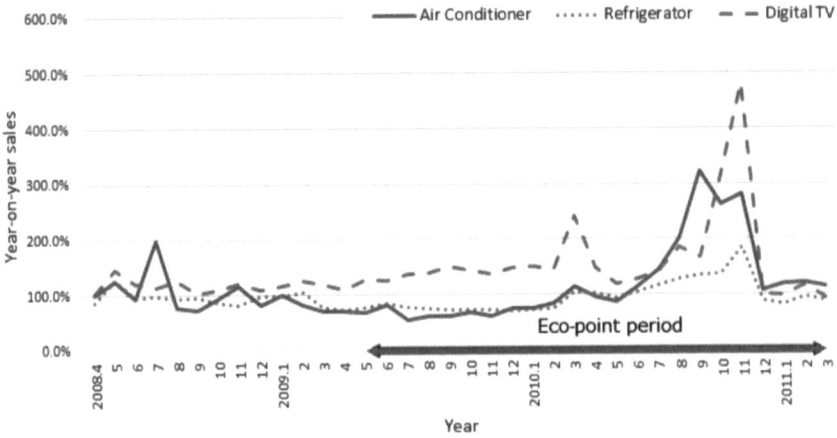

Figure 2.2 Historical change in sales of electric appliances

Source: Ministry of Economy, Trade, and Industry (2013)

2.5.3. *Eco-point program for housing*

The building and renovation of eco-friendly houses is intended to mitigate global warming problems as well as stimulate the country's economy. On 8 March 2010, the Japanese government introduced the eco-point program for housing to promote eco-friendly houses. The eco-point program for housing covers (1) new houses satisfying the top runner requirement or (2) new wooden houses satisfying the 2011 energy efficiency criterion.[6] It also provides subsidies for housing renovations to remove impediments for elderly and handicapped people and to improve the energy efficiency of houses though the installation of photovoltaic heat utilization systems.[7]

The program was initially planned for December 2009 to December 2010. However, there was a request to extend the program. The Japanese government compiled the contingent budget and extended the program another year. The budget, however, dried up in July 2011, and applications for the eco-point program for housing were closed at that time. On October 2011, the program was revived as a reconstruction support program and eco-point program for housing. This new program includes the recovery measure for the Great East Japan Earthquake.

The supplementary budget was compiled for 100 billion yen ($83.33 billion) in 2009. In the following year, a contingent budget of 141.2 billion yen ($1.18 billion) and supplementary budget of 3 billion yen ($25 million) were compiled. Finally, the supplementary budget was compiled for 144.6 billion yen ($1.21 billion) in 2011. Thus, the total budget of the eco-point program for housing was 388.8 billion yen ($3.24 billion) (Office of Housing Eco Point 2013).

In the initial eco-point program, a household building a new energy-efficient house received 300,000 yen ($2,500), while a household renovating a house received a subsidy for the renovations.[8] In the new eco-point program, the points were reduced to half, excluding the disaster-struck areas.

Eco points obtained could be used for construction costs or the expenses of other products. In addition to the provision of eco points, the government introduced an income tax reduction measure. In 2014, 10 percent of the construction cost (up to a maximum of 650,000 yen or $5,417) is deductible from income. Although 1.0 percent of mortgage balances are deductible from income for 10 years, borrowing for eco-friendly houses will be less limited than that for general houses (MLIT, 2012).

According to the statistics compiled by the Office of Housing Eco Point (2014), by March 2014, the number of applicants for the construction of new houses was 1,082,435, while that for the renovation of houses was 797,690. The amount of eco points issued for construction was worth 290.1 billion yen ($2.42 billion), and that for renovation was 51 billion yen ($0.43 billion).

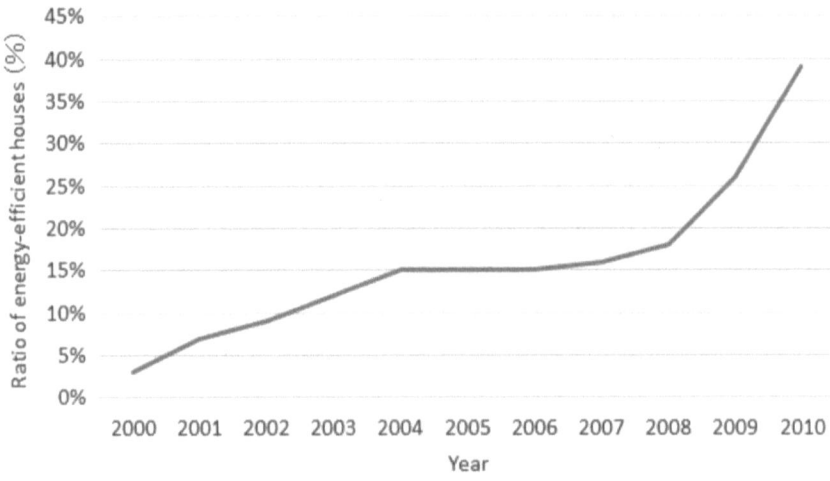

Figure 2.3 Historical change in the share of energy-efficient houses

Source: Ministry of Land, Infrastructure, Transport and Tourism (2013)

Figure 2.3 shows the historical change in the share of energy-efficient houses (MLIT 2013). The figure shows that the share of energy-efficient houses has steadily increased. Compared to the average price of new houses, 3,500,000 yen ($29,167) (MLIT 2013), the contribution of eco points amounting to 300,000 yen ($2,500) is relatively small. Nevertheless, the figure shows that energy-efficient houses became extremely popular after the introduction of the eco-point program for housing. Chino (2012) reports that 50 percent of new houses and 10 percent of renovations were part of the program in 2010.

2.5.4. Subsidy for photovoltaic generation systems

The Japanese government has introduced a wide variety of promotional measures for photovoltaic power generation systems (PPGSs) over the last two decades. The cost to install a PPGS of 3.5 kW designed for an average household was about 13 million yen ($108,333 in 1993. The government had introduced various subsidy programs for the installation of PPGSs from 1994 to 2005, attempting to reduce the manufacturing cost through mass production (Guide for the installation of photovoltaic generation systems for housing 2013). As a result of those subsidy programs, the installation cost of a PPGS was reduced to 2.3 million yen ($191,700), about one-sixth of the cost before the subsidy program.[9]

The series of PPGS subsidy programs were completed in 2005. A new subsidy program (Support Program for Installation of Photovoltaic Generation Systems for Houses) was introduced in 2008, and 104.1 billion yen ($867.5 million) were provided for the new subsidy program.

The amount of the subsidy was gradually reduced during the subsidy program period. Specifically, the rate of the subsidy was reduced from 70,000 yen/kW

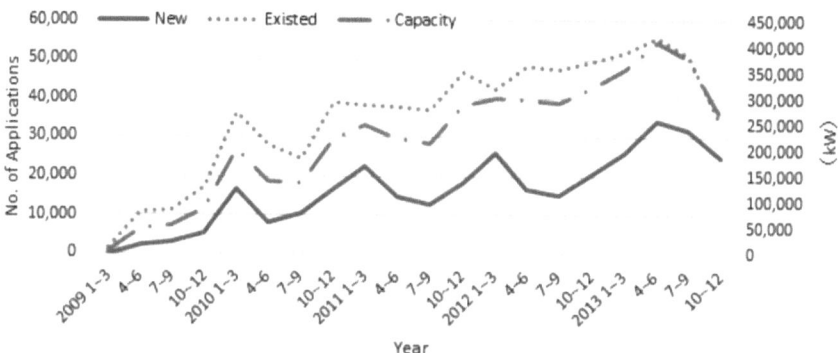

Figure 2.4 Historical change in support for the installation of photovoltaic power generation systems for houses

Source: Japan photovoltaic Expansion Center. (2014)

($583.33/kW) in 2009 to 20,000 yen/kW ($166.67/kW) in 2013. Consequently, the amount of subsidy that average households received was reduced from 250,000 yen ($2,083.33) to 70,000 yen ($583.33).

Figure 2.4 shows two different measures for the second program. The left axis measures the number of applications to install PPGSs in new and established houses and the right axis measures the installed capacity of the systems. The figure shows that the number of applications increased despite the reduction of the amount of the subsidy over the years of its implementation. The cumulative number of application for new houses was 326,442, while that for established houses was 708,305, on December 2013 (Japan Photovoltaic Expansion Center 2014).

In addition to the national government, local governments have provided a subsidy for the installation of PPGSs. Although the subsidy program by the national government concluded on March 2014, some local governments have continued their subsidy programs. The amount of the subsidy varies across local governments and ranges from 10,000 yen/kW ($83.33/kW) to 50,000 yen/kW ($416.67/kW).

The second major support for PPGSs is the Feed in Tariff (FIT) Scheme for Renewable Energy. In FIT, households can sell excess PV-generated electricity to a contracted utility company at a pre-specified price for 10 years. By selling excess electricity at a higher price, households can collect investment money within 10 years.

The purchase price of PV-generated electricity is decreasing. Specifically, the purchase price was 42 yen/kW ($0.35/kW) for PPGSs installed before 2012. The purchase price was reduced to 38 yen/kW ($0.32/kW) in 2013 and further reduced to 37 yen/kW in 2014.

The actual benefit depends on the type of contract with the utility company. At the Tokyo Electric Power Company, for example, the price of electricity is about 26–30 yen/kW ($0.22–0.25/kW), and a household can obtain 7–11 yen/kW ($0.06–0.09/kW) by selling PV-generated electricity to the company.

According to the statistics of the Agency of Natural Resources and Energy (2014), the cumulative capacity of PPGSs was 4.7 million kW on June 2012. Twenty-one months after the end of the FIT system, the cumulative capacity of PPGSs of 2.1 million kW were newly introduced.

2.5.5. Subsidy for garbage disposal apparatuses

Municipalities are responsible for the treatment of general waste and spend substantial amounts of money on waste management every year. According to MOE (2014) statistics, the average municipality spent 13,900 yen ($115.83) per person for the treatment of general waste in 2012. To lower the financial burden, all municipalities have implemented waste reduction measures. As part of these measures, many municipalities introduced a subsidy program for garbage disposal apparatuses.

According to a survey by the Japan Electrical Manufacturers' Association (2013), more than 60 percent of municipalities have provided subsidies for garbage disposal apparatuses.[10] Although the amount of the subsidies varies across municipalities, it averages around 50–60 percent of the purchase price or 20,000–50,000 yen ($166.67–416.67).

2.6. Conclusion

Environmental subsidies to consumers have become popular in recent years. This chapter explains the background behind their popularity and points out problems found in the implementation of subsidy programs in Japan. In this last section, we want to review the research agenda for environmental subsidy programs.

If a subsidy program is introduced to solve an environmental problem, it is necessary to examine whether the program will achieve its aim. Most subsidy programs have aimed at stimulating the country's economy, and past research has focused on analyzing the impact on sales expansion. At the current stage, the environmental impact of subsidy programs has not been fully examined. For instance, there is a diversity of opinion regarding the impact on CO_2 reduction of the eco-point program for appliances (MOE, MITE, Ministry of Public Management, Home Affairs, Posts, and Telecommunications [MPMHPT] 2011, Aoshima 2013).

If the mitigation of the environmental problem is confirmed, then we should discuss the efficiency of the program. To justify the implementation of the subsidy program, we have to show that it is more efficient than other environmental regulations. If the program is introduced for CO_2 reduction, then the unit costs to reduce CO_2 must be compared. Although there are few exceptions – such as Gillingham et al. (2009) – recent studies have not conducted such a calculation.

It is also necessary to examine whether the subsidy program generated an unintended effect/consequence. For instance, Kumano (2010) argues that subsidy programs, for a limited time, disrupt the sales cycle of durable products. This disruption occurs because the environmental subsidy program temporally relaxes the budget constraint of consumers. Consumers may purchase a large and energy-intensive product as a result of an environmental subsidy program. The examination of a rebound effect is therefore required (de Haan et al. 2006).

Although subsidy programs support the purchase of energy-efficient products, the amount of the subsidies depends on the size of the product. In general, the amount of the subsidy for a large product is large. This suggests that a rich household with an energy-intensive lifestyle will receive a larger benefit than a poor household with a less energy-intensive lifestyle. Hence, environmental subsidy programs can be regressive. Studies should examine what types of households use environmental subsidy programs.

In addition to the short-term impact, we should also evaluate the long-term impact of the program on pollution mitigation. For instance, we must evaluate how much energy consumption is reduced by the eco-point program for energy-efficient appliances or how much gasoline consumption was reduced by the eco-car subsidy.

In the current environmental subsidy programs, the environmental burden from product use is used as the environmental efficiency criteria. Environmental pollution generated at the production and disposal stages is mostly ignored. A reevaluation of environmental subsidy programs from a lifecycle point of view is needed. Also, while the potential environmental burden indicated in the product catalog is used as an environmental efficiency criterion, it is known that there is a large discrepancy between catalogued and actual environmental efficiencies. Perhaps a household survey is necessary to estimate the actual environmental burden.

Acknowledgment

This work is financially supported by JSPS KAKENHI Grant Number 26340119.

Notes

1 For example, the Water Pollution Control Law in Japan orders establishments to discharge water that does not meet effluent standards (MOE 2013).
2 Some scholars have examined whether consumers pay a premium for eco-friendly products that do not provide them a direct benefit. Teisl et al. (2002) found that Americans purchase tuna cans with the dolphin safe label – designed for the protection of dolphins – at a high price, and Bjorner et al. (2004) found that the Danish purchase toilet paper with the Nordic swan label – for the reduction of environmental pollution – at a high price. Unlike with energy-saving products, consumers are not directly benefitted by the consumption of these products. Therefore, these findings suggest that consumers pay extra purely for the purpose of environmental protection.
3 120 yen is converted into $1.
4 Most economic analyses use a discount rate of 4–6 percent.
5 A favorable depreciation rule is another common type of environmental subsidy, but it is often applicable only to firms and their investments. Therefore, since

such a rule is not common for consumers, we skip the discussion of it in this chapter.

6 In the top runner approach, the most efficient use of energy in primary electric machines is set as the standard in each industry.

7 The housing reform includes the replacement of windows and the installation of heat-insulating material, water-saving toilets, and bathtubs of a high-heat insulating specification.

8 The maximum amount of the subsidy for house renovation was 300,000 yen.

9 It is estimated that, as of 2014, the installation cost has since been reduced to 1.1–1.4 million yen.

10 There are 1,742 municipalities in Japan.

References

Agency of Natural Resources and Energy, (2014) 'Status of the installation of renewable energy facilities', Available: <http://www.meti.go.jp/press/2013/01/2014 0110002/20140110002.html> (accessed 14 April 2014) (In Japanese).

S. Anderson and R. Newell, (2004) 'Information programs for technology adoption: The case of energy efficiency audits', *Resource and Energy Economics* 26, 27–50.

Y. Aoshima, (2013.8.29) 'Ask the evaluation eco-point program for appliances and feed in tariff system for renewable energy', *Weekly Keidanren Times*. (In Japanese.)

T.H. Arimura, S. Li, R.G. Newell and K. Palmer, (2012) 'Cost-effectiveness of electricity energy efficiency programs', *The Energy Journal* 33, 63–99.

M. Auffhammer, C. Blumstein and M. Fowlie, (2008) 'Demand-side management and energy efficiency revisited', *The Energy Journal* 29, 91–104.

Automobile Information Center, (2013) 'Time trend of the number of registered hybrid vehicles', Available: <http://autoinfoc.com> (accessed 12 April 2014).

W.J. Baumol and W.E. Oates, (1971) 'The use of standards and prices for protection of the environment', *The Swedish Journal of Economics* 73, 42–54.

W.J. Baumol and W.E. Oates, (1988) *The Theory of Environmental Policy*, 2nd edition. Cambridge University Press.

T.N. Bjorner, L.G. Hansen and C.S. Russell, (2004) 'Environmental labeling and consumers' choice – An empirical analysis of the effect of the Nordic Swan', *Journal of Environmental Economics and Management* 47, 411–34.

J.A. Caswell and E.M. Mojduszka, (1996) 'Using informational labeling to influence the market for quality in food products', *American Journal of Agricultural Economics* 78, 1248–53.

T. Chino, (2012.2.22) 'Evaluation of eco-policies for housing', *Mizuho Report*. Singapore, Mizuho Research Institute.

P. De Haan, A. Peters and R.W. Scholzm, (2006) 'Reducing energy consumption in road transport through hybrid vehicles: Investigation of rebound effects, and possible effects of tax rebates', *Journal of Cleaner Production* 15, 1077–84.

K.M. Gallagher and E. Muehlegger, (2011) 'Giving green to get green? Incentives and consumer adoption of hybrid vehicle technology', *Journal of Environmental Economics and Management* 61, 1–15.

K. Gillingham, R.G. Newell and K. Palmer, (2009) 'Energy efficiency economics and policy', Discussion paper RFF-DP 09–13, Washington, DC, Resources for the Future.

Guide for the installation of photovoltaic generation systems for housing, (2013.12.5) 'The deadline for the application of subsidy is March 31, 2014', Available: <http://

www.qool-shop.com/topix/entry109.html> (accessed 23 April 2014) (In Japanese).

K.A. Hassett and G.E. Metcalf, (1995) 'Energy tax credits and residential conservation investment: Evidence from panel data', *Journal of Public Economics* 57, 201–17.

E. Hosoda, (2000) 'Asymmetry of control and quantity control in an environmental policy', *Environmental Economics and Policy Studies* 3, 381–97.

A. Jaffe, R. Stavins and R. Newell, (1995) 'Dynamic incentives of environmental regulations: The effects of alternative policy instruments on technology diffusion', *Journal of Environmental Economics and Management* 29, S43–S63.

Japan Electrical Manufacturers' Association, (2013) 'Garbage disposal apparatuses for households', Available: <http://www.jema-net.or.jp/Japanese/ha/gomi/grant.html> (accessed 23 April 23 2014).

Japan Photovoltaic Expansion Center, (2014) Press release, Available: <http://www.j-pec.or.jp/information/press.html> (accessed 23 April 2014).

T. Kawase, (2011) 'Country support under world financial depression and WTO subsidy rule', *Research Institute of Economy, Trade & Industry.* RIETI Discussion Paper Series 11-J-065. (In Japanese.)

H. Kumano, (2010.11.30) 'Rebound sales reduction after eco-car subsidy and eco-point program: The magnitude of pre purchase varies among vehicles and appliances', Economic Trend, Dai-Ichi Life Research Institute, Inc., Available: <http://group.dai-ichi-life.co.jp/dlri/kuma/pdf/k_1012a.pdf> (accessed 14 April 2014) (In Japanese).

D. Loughran and K. Kulick, (2004) 'Demand-side management and energy efficiency in the United States', *The Energy Journal* 25, 19–41.

S. Matsumoto and T. Yokoyama, (2008) 'Discriminative treatment in environmental regulation', in Y. Akira and Ministry of Finance, Policy Research Institute (Eds), *Global Warming Measure and Policy Design for Economic Growth,* Chapter 8. Keiso Shobo, 207–29.

METI, (2012.9.21) 'Eco-car subsidy', Available: <http://www.meti.go.jp/topic/data/091112aj.html> (accessed 13 April 2014) (In Japanese).

METI, (2013) 'Monthly sales statistics of home electric appliances at volume sellers', Available: <http://www.meti.go.jp/statistics/zyo/kadenhan> (accessed 4 March 2014) (In Japanese).

MLIT, (2012) Announcement about taxation about housing taxation, Available: <http://www.mlit.go.jp/common/001035432.pdf> (accessed 13 April 2014) (In Japanese).

MLIT, (2013) 'Energy saving measures for house and building', Available: <http://www.mlit.go.jp/common/000193924.pdf> (accessed 13 April 2014) (In Japanese).

MOE, (1980) 'Chapter 3, Improvement of environmental policies', White Paper on the Environment, Available: <http://www.env.go.jp/policy/hakusyo/past_index.html> (accessed 5 April 2014) (In Japanese).

MOE, (1998) 'Chapter 4, Promotion of the basic measures commonly associated with environmental conservation', white paper on the environment, Available: <http://www.env.go.jp/policy/hakusyo/past_index.html> (accessed 5 April 2014) (In Japanese).

MOE, (2012.8.31) 'Announcement: Status of the promotion program of green electric appliances with the application of eco points', Available: <https://www.env.go.jp/policy/ep_kaden/about/report.html> (accessed 12 April 2014) (In Japanese).

MOE, (2013) *Water Pollution Control Act*, Available: <http://law.e-gov.go.jp/htmldata/S45/S45HO138.html> (accessed 24 March 2014) (In Japanese).

MOE, (2014). *Waste Management of Japan*, Available: <http://www.env.go.jp/recycle/waste_tech/ippan/h24/data/disposal.pdf> (accessed 8 May 2014) (In Japanese).

MOE, MITE, and MPMHPT, (2011) 'About the policy impact of eco-point programs for electric appliances', Available: <http://www.meti.go.jp/press/2011/06/20110614002/20110614002-2.pdf> (accessed 14 April 2014) (In Japanese).

OECD, (1972.5.26) 'Recommendation of the council on guiding principles concerning international economic aspects of environmental policies', C(72)128, Available: <http://webnet.oecd.org/OECDACTS/Instruments/ShowInstrumentView.aspx?InstrumentID=4&InstrumentPID=255&Lang=en&Book> (accessed 5 March 2014).

OECD, (1974.11.14) 'Recommendation of the council on the implementation of the polluter pays principle', C(74)223, Available: <http://acts.oecd.org/Instruments/ShowInstrumentView.aspx?InstrumentID=11> (accessed 5 March 2014).

OECD, (2011). *Greening Household Behaviour: The Role of Public Policy*. OECD Publishing.

Office of Housing Eco Point, (2013) 'Eco point for housing', Available: <http://jutaku.eco-points.jp/user/outline> (accessed 22 April 2014).

Office of Housing Eco Point, (2014) 'Status of eco point for housing', Available: <http://fukko-jutaku.eco-points.jp/pressrelease/140418_1.html> (accessed 23 April 2014).

D. Popp, (2002) 'Induced innovation and energy prices', *American Economic Review* 92, 160–180.

A.H. Sanstad, W.M. Hanemann and M. Auffhammer, (2006) 'End-use energy efficiency in a "post-carbon" California economy: Policy issues and research frontiers', in M.W. Hanemann and A.E. Farrell (Eds), *Managing Greenhouse Gas Emissions in California*, Chapter 6. The California Climate Change Center at UC Berkeley.

R.N. Stavins, (2003) 'Experience with market-based environmental policy instruments', in K.G. Mäller and J.R. Vincent (Eds), *Handbook of Environmental Economics Vol.1, Environmental Degradation and Institutional Reponses*, Chapter 9. North Holland.

M.F. Teisl, B. Roe, and R.L. Hicks, (2002) 'Can eco-labels tune a market? Evidence from dolphin-safe labeling', *Journal of Environmental Economics and Management* 43, 339–59.

U.S. Department of Commerce, (2014) 'Commerce preliminarily finds countervailable subsidization of imports of certain crystalline silicon photovoltaic products from the People's Republic of China', Available: <http://enforcement.trade.gov/download/factsheets/factsheet-prc-crystalline-silicon-photovoltaic-prod-cvd-prelim-060314.pdf> (accessed 23 July 2014).

WTO, (2014) 'Uruguay round agreement, Agreement on Subsidies and Countervailing Measures', Available: <http://www.wto.org/english/docs_e/legal_e/24-scm_01_e.htm> (accessed 23 July 2014).

3 Environmental tax and subsidy in Japan

Past and present

Satoshi Sekiguchi

3.1. Introduction

In general, the discussion of environmental policy measures has focused on direct regulations and economic measures (such as subsidies, green taxes, and fees). This chapter focuses on the classification of financial and non-financial measures. Public finance actions are policy measures that give rise to government spending and revenue. Non-public finance actions are policy measures such as direct regulations that do not bring about government spending and revenue.

There are two different ways that environmental policy can be carried out according to public finance actions. First, central or local governments can carry out environmental conservation measures directly. Second, such measures can be carried out indirectly through the conservation actions of other parties.[1]

When the central government or local governments utilize their annual expenditure to implement environmental policies, they provide funds directly to environmental conservation projects. In addition, policy objectives can be carried out indirectly by means of providing subsidies to consumers and manufacturers. Furthermore, when the central government or local governments utilize their tax system to implement environmental policies, the policy objectives are met via tax reductions or increases on the manufacturers and consumers.

Chapter 1 discussed consumer subsidies such as favorable interest rates, sales discounts, and tax reduction measures. Favorable interest rates and sales discounts are policies that utilize the annual expenditure of the central and local governments. Tax reduction measures are policies that utilize annual revenue.[2]

This chapter provides a historical review of environmental taxes/subsidies that have been implemented in Japan. The chapter compares national and local policies and summarizes the problems that policymakers faced when introducing environmental taxes/subsidies.

3.2. The historical development of Japan's environmental public finance

This section will provide a quantitative understanding of the history and idiosyncrasies of environmentally motivated government spending and taxation policy.

"Environmental public finance" refers to public finance actions related to the environment that result in government revenue and spending.

3.2.1. The era of industrial pollution

Following World War II, the Japanese government implemented industrial policies that were centered on the heavy chemical industry. The goals of these policies were to strengthen the international competitiveness of Japanese companies through the development of large-scale manufacturing facilities and a reduction in manufacturing costs. Thus, pollution-control measures were put off. The result was an increase in industrial pollution, such as soil, water, and air pollution. This set up a situation where there was confrontation between "industry" and "local residents." In order to deal with industrial pollution, the government implemented environmental policies in the form of directly regulating pollutants emitted by companies' manufacturing activities.

3.2.1.1. The historical transition of environmentally motivated expenditure policy

Environmental conservation expenditures (expenses related to the prevention of pollution and the protection and maintenance of the environment) during the era of industrial pollution are summarized in Table 3.1.

First, Table 3.1 confirms the rapid expansion of environmental conservation expenses by the central government during the 1970s. The amount of central government expenditure for environmental conservation in 1971 was the exceedingly small sum of 111.4 billion yen. While it is essential to account for inflation, this number increased 10-fold, to 1.1664 trillion yen, by 1980. The increase in the central government's environmental conservation expenses can be confirmed in the percentage of environmental conservation expenses that comprise the central government's total expenditure. In 1971 it was no more than 0.68 percent, but it increased by 0.81 percent, to 1.49 percent, in 1980.

Second, it can be confirmed that local governments' expenses for anti-pollution measures were larger than those of central government in the 1970s. In 1971, local governments spent 586.6 billion yen on anti-pollution measures, roughly five times that spent by the central government (111.4 billion yen). In 1980, the amount of local governments' expenses for anti-pollution measures increased to 2.7514 trillion yen. However, because the amount of the central government's environmental conservation expenses increased more than this, the ratio of the local governments' environmental conservation expenses to the central government's environmental conservation expenses was reduced from 5 times to roughly 2.6 times.

The increase in the local government's environmental conservation expenses can be confirmed in the percentage of environmental conservation expenses that comprised total local government expenditure. In 1971, it was 4.93 percent, but it increased by 1.08 percent to 6.01 percent in 1980. As will be shown later, it is assumed that the increase in the percentage of expenditures for anti-pollution

Table 3.1 Environmental expenditures of the central government and local governments
(unit: JPY 100 million, %)

Year	Central Government[a]				Local Governments[b]		
	Total Expenditure (net) ①	Environmental Conservation Expenditure[c,d,e,f] ②	Subsidies to Local Governments	②/①	Total Expenditure (net) ③	Pollution-Control Expenditure ④	④/③
1971	163,936	1,114	n.a	0.68	119,095	5,866	4.93
1972	193,616	1,693	n.a	0.87	146,182	8,113	5.55
1975	358,977	3,751	n.a	1.04	256,544	14,258	5.56
1980	781,129	11,664	n.a	1.49	457,807	27,514	6.01
1985	971,029	11,172	n.a	1.15	562,934	27,568	4.90
1990	1,218,562	13,402	n.a	1.10	784,732	37,218	4.74
1995	1,595,419	25,987	19,328	1.63	989,445	61,195	6.18
2000	2,088,092	30,420	n.a	1.46	976,163	55,617	5.70 ·
2005	2,396,553	23,654	n.a	0.99	906,973	32,198	3.55
2009	2,065,158	21,168	n.a	1.03	961,064	25,850	2.69
2010	2,150,656	12,596	n.a	0.59	947,750	23,180	2.45

a The amount initially budgeted by the central government. The duplication of subsidies between general account and special accounts etc. in central government is deducted from the combined expenditure of general account and special accounts etc. in central government. Subsidies to local governments are the expenses handed over to local governments and monetary subsidies that local governments use for specific purposes.

b Due to limitations of the data, the local government amounts are actual value. The duplication of subsidies between municipalities and prefectures are deducted from the combined expenditure of municipalities and prefectures. Local governments' pollution-control expenditures do not include expenses for natural environmental conservation and global warming measures. In short, these figures are more limited than the range of the central government's environmental conservation expenditure.

c Regarding the Environmental Conservation Budget for FY1994, the scope was expanded and relevant expenditures were allocated to comply with the Basic Environment Plan, which was formulated under the provisions of the Basic Environment Law in 1994.

d Regarding on the Environmental Conservation Budget for FY2000, the relevant expenditures were allocated, excluding the reduced adjusted expenditure posed by the establishment of independent administration institutes from FY2001.

e Expenditures related to the promotion of locating nuclear power plants was allocated to the Environmental Conservation Budget from FY2008.

f As for the new subsidy relevant to the public works–related expenditures implemented in FY2010, the use is not specified. Such costs are not allocated to the Environmental Conservation Budget, and thus cannot be compared and contrasted with the previous fiscal year.

Source: Ministry of Environment and Ministry of Internal Affairs and Communications (various issues).

measures that comprise the local government's total expenditure was influenced by subsidies from the central government.

In order to deal with industrial pollution, environmental policies during this time focused on the central government and local governments paying directly for environmental conservation projects and taking measures to assist those companies that were subject to the new legal regulations. The local governments led the way on implementing legal regulations until the Environmental Pollution Prevention Act was passed by the central government. In a similar manner, it has been pointed out that the environmental conservation expenses of local governments at the beginning of the 1970s were larger relative to the environmental conservation expenses of the central government.

3.2.1.2. The historical transition of environmentally motivated taxation policy

One universal characteristic of taxation policy in the era of industrial pollution was with the use of both central and local legal regulations to deal with industrial pollution and natural conservation. These worked together to provide tax reduction (tax breaks) through taxation policy to impel manufacturers to reduce their production of pollutants. The particular characteristics of this taxation policy are shown in Table 3.2.

First, tax reductions were incorporated as part of the larger taxation policy. These tax reductions were targeted primarily at companies that were involved in manufacturing. Specifically, corporation taxes were postponed via a special repayment for facilities that prevented pollution: a decrease in the fixed assets tax for waste treatment facilities (~1972) and a decrease in the fixed assets tax on facilities that processed sewage and liquid waste (~1976). The goal in reducing (or postponing) the corporation tax (a national tax) and the fixed assets tax (a local tax) was to reduce the burden on companies that were investing in anti-pollution measures.

In short, companies involved in manufacturing were legally required to comply with anti-pollution measures but the central government and the local governments worked to reduce the burden on companies carrying out these investments by reducing their tax burden. The characteristic point here is that there was a fundamentally uniform policy taken nationwide with regard to both national and local taxes.

Second, awareness of the so-called environmentally related taxes (a tax system that gives consideration to the environment) was exceedingly low at that time. The characteristics of the Japanese taxation system become clear in Table 3.2, which classifies the various Japanese taxes according to the OECD's environmentally related tax categories.[3]

The majority of Japan's environmentally related taxes were established during the period of rapid economic growth (1954–1973) when industrial pollution became a problem. However, when these taxes were initially established, they were for the purpose of securing financial resources for infrastructure improvement. The tax revenue from these taxes was set aside to support road maintenance (gasoline tax, Liquefied petroleum gas tax, automobile weight tax, diesel oil delivery tax, automobile acquisition tax), airport maintenance (aviation fuel tax), the stable

Table 3.2 Environmentally related taxes[c] (Unit: JPY 100 million)

Tax Item[a]	Founded Year	Type of Tax	Level of Tax	Use of Revenue	Main Ministries[b]	1970	1975	1980	1985	1990	1995	2000	2005	2009	2011
Gasoline tax[d]	1949	E	N	Road maintenance and improvement	Ministry of Land, Infrastructure and Transport (MLIT)	4,987	8,244	15,474	16,678	20,066	24,627	27,686	29,084	27,152	26,484
Liquefied Petroleum gas tax	1966	E	N			122	139	149	155	157	153	142	142	123	113
Automobile weight tax	1971	M	N			–	2,203	3,951	4,523	6,609	7,837	8,507	7,574	6,351	4,478
Petroleum and coal tax	1978	E	N	Fuel stable supply	Ministry of Economy, Trade and Industry (METI)	–	–	4,041	4,004	4,870	5,131	4,890	4,931	4,868	5,191
	2003	E	N	Sophistication of the energy supply and demand structure	Ministry of the Environment (MOE)	–	–	–	–	–	–	–			
Aviation fuel tax	1972	E	N	Airport maintenance and improvement	Ministry of Land, Infrastructure and Transport (MLIT)	–	183	488	521	641	855	880	886	793	462

(Continued)

Table 3.2 (Continued)

Tax Item[a]	Founded Year	Type of Tax	Level of Tax	Use of Revenue	Main Ministries[b]	1970	1975	1980	1985	1990	1995	2000	2005	2009	2011
Promotion of power resources development tax	1974	E	N	Support to the construction and operation of power generation facilities	Ministry of Economy, Trade and Industry (METI)	—	299	1,085	2,335	2,947	3,386	3,746	3,592	3,293	3,314
	1980	E	N	Safety and promoting the use of power generation facilities	Minister of Education, Culture, Sports, Science and Technology (MEXT)	—	—								
Local gasoline transfer tax	1955	E	L	Road maintenance and improvement	Ministry of Land, Infrastructure and Transport (MLIT)	903	1,496	2,783	2,999	3,608	2,635	2,962	3,112	2,905	2,834
Light oil delivery tax	1956	E	L			1,442	1,940	4,471	5,558	8,335	13,322	12,076	10,859	9,084	9,318
Petroleum gas transfer tax	1966	E	L			122	139	149	155	157	153	142	142	123	113
Automobile weight transfer tax	1971	M	L			—	734	1,317	1,508	2,203	2,612	2,836	3,787	3,176	3,073
Automobile acquisition tax	1968	M	L			764	1,750	2,703	3,471	6,131	6,112	4,641	4,528	2,310	1,678

Item	Year	Class[b]	Class[b]	Special account	Ministry[b]										
Automobile tax[c]	1950	M	L	—	—	1,714	3,689	7,806	10,380	12,762	15,873	17,645	17,528	16,544	15,972
Light motor vehicle tax[f]	1958	M	L	—	—	238	275	432	698	881	1,055	1,249	1,515	1,739	1,804
Aviation fuel transfer tax	1972	E	L	Airport maintenance and improvement	Ministry of Land, Infrastructure and Transport (MLIT)	—	33	89	95	116	155	160	161	144	132
Total amounts ①						10,292	21,124	44,938	53,080	69,483	83,906	87,562	87,84	78,605	74,966
Revenue for road maintenance and improvement ②						8,340	16,645	30,997	35,047	47,266	57,451	58,992	59,228	51,224	48,091
Share of the total amounts ②/①						81.0%	78.8%	69.0%	66.0%	68.0%	68.5%	67.4%	67.4%	65.2%	64.2%
E: Tax on energy products						7,576	12,473	28,729	32,500	40,897	50,417	52,684	52,909	48,485	47,961
M: Tax on motor vehicles and transport						2,716	8,651	16,209	20,580	28,586	33,489	34,878	34,932	30,120	27,005
N: Environmentally related tax revenue of national government						5,109	11,068	25,188	28,216	35,290	41,989	45,851	46,209	42,580	40,042
Share of the total national tax						6.6%	7.6%	8.9%	7.2%	5.6%	7.6%	8.7%	8.8%	10.6%	8.9%
L: Environmentally related tax revenue of local government						5,183	10,023	19,661	24,769	34,077	41,762	41,551	41,471	35,881	34,792
Share of the total local tax						13.8%	12.3%	12.4%	10.6%	10.2%	12.4%	11.7%	11.9%	10.2%	10.2%

a Because of a minor change from 1970, tax items use the name of the 2014 point.
b Major ministries have been classified on the basis of the authorities, etc. that have jurisdiction over the special account associated with a particular financial resources.
c In addition to this, there is an industrial waste tax (a local tax). This tax was 6.9 billion yen in 2012.
d The old gasoline tax was established in 1937 and then abolished in 1943. It was revived in 1949. Revenue from the gasoline tax was earmarked for road construction in 1954.
e The old automobile tax was introduced in 1930. The current automobile tax was established in 1950.
f The bicycle tax, a forerunner of the light vehicle tax, was established in 1940.

Source: Author's calculation based on data from the Policy Research Institute, Ministry of Finance.

provision of fuel (petroleum tax), and the promotion of power generation (promotion of power resources development tax).

This meant that the ministries and government offices that these tax revenues flowed to had the authority to spend them as they saw fit. The two most influential ministries were the Ministry of Land Infrastructure, Transport and Tourism (formerly the Ministry of Construction) and the Ministry of Economy, Trade and Industry (formerly the Ministry of International Trade and Industry). Although these specific financial resources were useful in improving Japan's infrastructure, they were an underlying cause of lower civic awareness of Japan's environmentally related taxes.

For example, when one looks at how the revenue generated by environmentally related taxes is divided, one can see that the central government and the local governments divide these taxes in a similar manner. However, there is little awareness of carrying out environmental policy when these revenues are acquired, and there is the persistent belief that these financial resources should be used for infrastructure improvement (road maintenance etc.).

3.2.2. *The period of household pollution*

By the beginning of the 1980s, pollution caused by industrial manufacturing showed signs of stabilizing. However, there was a striking increase in lifestyle and urban pollution (household pollution) that went along with everyday life and daily activities. This was due to the concentration of the population in large urban areas and a change in lifestyle. Increased urbanization was accompanied by an increase in air pollution due to the pollutants emitted from vehicles and water pollution due to wastewater.[4] This was significant because it gave rise to problems that could not be handled using the existing formula of individually enumerating the industrial activities that caused the environmental burden and regulating them. In short, the confrontation between "manufacturing" and "local residents" had changed.

3.2.2.1. *The historical transition of environmentally motivated expenditure policy*

Environmental conservation expenses began to have the following characteristics in the household pollution period due to the emergence of issues such as waste management and air pollution caused by automobile emissions.

First, the central government's environmental conservation expenditure as a percentage of total expenditure decreased temporarily after the 1980s, but then increased in the latter half of the 1990s. It increased from 1.1 percent in 1990 to 1.63 percent in 1995, due, in part, to the influence of Basic Environment Law of 1993 that expanded the range of targets of environmental conservation spending (supra Table 3.1).

However, the percentage of environmental conservation expenditure that comprised the central government's expenditure subsequently decreased; it was 0.99 percent in 2005 and 1.03 percent in 2009. This trend can also be confirmed when looking at the decrease in funds available for environmental conservation expenses. The primary causes for this can be verified by paying attention to the fields covered by the central government's environmental policies (Table 3.3).

Table 3.3 Environmental conservation expenditures[a] by the central government (unit: %, JPY 100 million)

	1972	1975	1980	1985	1990	1995	2000	2005	2009	2010
Realization to the recycling economy and society[b,c]	88.0	86.4	89.7	89.1	87.1	84.5	77.8	82.1	84.0	81.6
Conservation of the global environment[e]	n.a.	n.a.	n.a.	n.a.	n.a.	n.a.	20.6	23.0	32.0	49.2
Conservation of the atmospheric environment[e]	n.a.	n.a.	n.a.	n.a.	n.a.	n.a.	6.7	13.1	11.1	16.8
Conservation of water environment, soil environment, and ground environment[e]	n.a.	n.a.	n.a.	n.a.	n.a.	n.a.	43.7	39.1	35.1	8.1
Waste and recycling measures[c]	n.a.	n.a.	n.a.	n.a.	n.a.	n.a.	6.3	6.3	5.4	6.8
Chemical substance control[c]	n.a.	n.a.	n.a.	n.a.	n.a.	n.a.	0.6	0.6	0.4	0.6
Ensure the coexistence of man and nature[d]	10.0	11.3	9.1	9.8	11.1	12.6	20.0	14.1	12.3	11.7
Others	2.0	2.3	1.2	1.1	1.8	2.9	2.2	3.8	3.7	6.7
Total (%)	100.0	100.0	100.0	100.0	100.0	100.0	100.0	100.0	100.0	100.0
Total (JPY 100 million)	1,693	3,753	11,663	11,172	13,402	25,986	30,419	23,604	21,168	12,595

a Because categories were added or changed between 1972–1993, 1994–1999, and 2000–2013, the number of categories within each range are not contiguous.
b The items "realization to the recycling economy and society," "ensure the coexistence of man and nature," and "others" were used from 1994 to 1999.
c The item "pollution control measures," used from 1972 to 1990, was classified into "realization to the recycling economy and society."
d The item "conservation of natural environment," used from 1972 to 1990, was classified into "ensure the coexistence of man and nature."
e The item "realization to the recycling economy and society" was subdivided in 2000.

Source: Author's calculations based on data from the Ministry of Environment.

Unfortunately, for dates prior to 1990, a clear breakdown of the fields covered by these policies is available. However, by looking at other materials, it is clear that the field that was covered by policy funded from 1972 and through the 1990s by the central government's environmental conservation expenditure, and comprising the majority (75 to 85 percent) of this expenditure, was consistently the "prevention of pollution related to public industry."[5]

After 2000, a more detailed breakdown becomes clear. At a glance, it becomes apparent that while there was an increase in expenditure related to "conservation of the global environment," there was a decrease in expenditure related to "conservation of water, soil, and land environments." After the turn of the millennium, there was an expansion of expenditure related to the "conservation of the global environment" in order to deal with the newly important political issue of global warming. The other side of this coin was that in order to prevent the budget deficit from increasing, public works spending was decreased. Thus, there was a reduction in expenditure related to the "conservation of water, soil, and land environments" that had been implemented for public utilities.[6]

Second, the percentage of expenditure for anti-pollution measures that comprised the total expenditure of local governments temporarily decreased after the 1980s, but increased from 4.74 percent in 1990 to 6.18 percent in 1995 (supra Table 3.1).

However, the percentage of expenditure for anti-pollution measures that comprised the total expenditure of local governments saw a sharp decrease to 2.69 percent in 2009. This trend can be confirmed by looking at the decrease in the funds available for expenditures on anti-pollution measures. The main causes for this will be verified through the amount spent on anti-pollution measures by the local governments by field (Table 3.4).

One characteristic of local government expenditures for anti-pollution measures is that "pollution prevention projects" (in particular, the maintenance of sewage projects and waste processing facilities) have comprised 60 to 70 percent of total expenditures. This has been historically consistent.

There was no change in the ratio for this trend in the 2000s. But when we compare it with the amounts in 2005 and 2009, there is a striking decrease in amounts available for anti-pollution expenses.[7] With regards to this, the OECD (2010) has pointed out that local governments are increasingly shifting waste matter processing to the private sector. One additional point that should be clarified is the scale of the subsidiary aid for localities that comprises the central government's environmental conservation expenses.

Due to the limitations of the data, only a limited number of years can be examined, but within the central government's environmental conservation expenses of 2.5987 trillion yen in 1995, the amount allocated to subsidies for localities was 1.9328 trillion yen,[8] or 74.37 percent of the central government's environmental conservation expenses. In 1997, the percentage of the central government's environmental conservation expenses allocated as subsidies for localities was 76.09 percent, and it increased to 96.12 percent in 1998.[9]

This suggests that while the local governments are at the core of implementing environmental policy, the majority of the money available for environmental measures is provided by the central government. In short, this means that when

Table 3.4 Pollution control expenditures[a] by local governments (unit: %, JPY 100 million)

	1972	1975	1980	1985	1990	1995	2000	2005	2009	2010
General expenses[b]	5.2	4.8	3.7	3.5	3.3	2.8	3.4	3.0	7.1	8.0
Pollution control and research expenditures	n.a.	n.a.	n.a.	1.1	1.2	0.8	0.9	0.9	1.6	1.5
Pollution prevention project cost[c]	91.1	89.6	90.2	88.9	90.4	92.8	93.2	94.1	87.0	83.7
Sewerage works	62.8	59.6	68.6	68.6	71.8	71.7	72.1	74.7	70.1	67.1
Development of waste processing facilities	14.6	17.5	11.4	12.1	12.1	16.0	17.3	15.2	12.1	12.0
Damage compensation expense of pollution health insurance	0.1	1.1	2.6	3.7	2.9	1.6	1.5	1.5	2.4	2.7
Others	3.5	4.5	3.5	2.8	2.2	2.1	0.7	0.6	1.8	4.1
Total (%)	100.0	100.0	100.0	100.0	100.0	100.0	100.0	100.0	100.0	100.0
Total (JPY 100 million)	8,113	14,258	27,514	27,569	37,218	61,195	55,167	59,687	25,850	23,180

a The duplication of subsidies, contributions, etc. between municipalities and prefectures are deducted from the combined expenditure of municipalities and prefectures. Pollution control expenditures of local governments do not include expenses for natural environmental conservation measures and global warming measures. In short, these figures are more limited than the range of environmental conservation expenditure by the central government.
b The item "ordinary expenses," used from 1972 to 1982, was classified into "general expenses."
c The item "construction project cost," used from 1972 to 1982, was classified into "pollution prevention project cost."

Source: Ministry of Internal Affairs and Communications (various issues).

there is a change in the amount of money and the standard for subsidies for localities that are related to environmental conservation by the central government, the amount of money for environmental conservation expenditure by the local governments is impacted. In recent years, the reduction of public works spending by the central government has had an impact in the form of a reduction in subsidies to local governments. The result of this has been a reduction in spending on anti-pollution measures by the local governments.

3.2.2.2. *The historical transition of environmentally motivated taxation policy*

A universal characteristic of the taxation policy in the period of household pollution (when sewage and air pollution due to automobile emissions became a problem) was that, in addition to the extension of tax breaks, discussions emerged on tax policies directed at the new problem of global warming. After 2000, there was a shift within the taxation system whereby environmental policy became linked with national and local taxes. This shift was from a complete devotion to reducing manufacturers' tax burdens to redesigning the existing taxation system and introducing new taxes (supra Table 3.2). The particular features of this are described below.

First, there was a change in the taxation system related to automobiles. The central government sought to incorporate environmental considerations into the tax code in an attempt to turn the so-called automobile tax into something greener. In 2001, there were tax breaks to subsidize environmentally friendly automobiles when the automobile tax (a general financial resource for local governments) was taken. The main feature of these considerations was that the automobile tax was not a financial resource that should be exclusively set aside for road construction by local governments, but rather that it should be treated as a general financial resource. In 2009, there were tax breaks to subsidize environmentally friendly automobiles when the automobile weight tax (national tax) and the automobile acquisition tax (local tax) were taken. These tax breaks were named the eco-car tax reductions.[10] The primary feature of both of these automobile-related taxes is that until 2008 they were financial resources that were exclusively set aside for road construction; in 2009 they became general financial resources.

Second, a new national tax was implemented by the central government in an effort to address global warming. After 2012, a tax targeted at global warming was implemented, in addition to the existing petroleum and coal tax. This was the first tax implemented with the intention of controlling carbon dioxide emissions. The tax revenue generated by this tax was limited to fund measures to combat global warming.

Third, the local governments independently introduced a tax on industrial waste (2001) and a green forest tax (2003). These taxes are not the same nationwide, because they were independently introduced by each local government, and the tax revenue that they generate is restricted for use with environmental policy.[11] These local taxes comprise an extremely small percentage of local tax revenue.[12] Note that the green forest tax does not meet the OECD's definition of an environmentally related tax by being a "tax on objects of taxation that are thought to be related to the environment." This is because it uses income and the number of people as its tax base.

3.3. The current state of Japan's environmental policy and the Ministry of the Environment

This section will take a look at the current state of environmental policy and the position of the Ministry of the Environment (MOE) in the formation of environmental policy. In the policymaking process, it is not just the administrative process but also the political process that is important. Thus, this section will pay attention to the interactions between the various administrative organizations.

3.3.1. Environmental expenditure policy in the political sphere

3.3.1.1. The formation of environmental expenditure policy and the Ministry of the Environment

The Environmental Pollution Prevention Act was enacted in 1967 to address industrial pollution. However, responsibility for pollution administration at that time was divided between a number of different ministries: the Ministry of Health and Welfare (now the Ministry of Health, Labour and Welfare), the Ministry of International Trade and Industry (now the Ministry of Economy, Trade and Industry), and the Ministry of Construction (now the Ministry of Land, Infrastructure, Transport and Tourism). Each of these ministries had different responsibilities with regards to pollution administration, making it difficult to implement pollution administration in a comprehensive and preventative manner.[13] Thus, the Environment Agency (now the MOE) was established in 1971 as a government office to handle the comprehensive coordination of pollution regulations and the concerned government agencies. It did this by integrating the various departments related to the environment that were dispersed across each ministry.[14]

However, it was not the case that the Environment Agency alone became the government office with jurisdiction over the entirety of environmental policy. The Environment Agency was merely given the power to comprehensively coordinate the business related to the conservation of the environment by the concerned administrative organs. For example, with regards to the environmental conservation budget, the Environment Agency adjusted the guidelines for giving a quote when the concerned government agencies requested a budget related to the environment. However, if one looks at the percentage of money that was demanded by government agencies for environmental conservation from the 1970s to the 1980s, the Environment Agency (now the MOE) was around 3 to 5 percent. The Ministry of Construction and the Ministry of Transport (now the Ministry of Land, Infrastructure, Transport and Tourism) comprised 65 to 75 percent of the budget (Table 3.5).

One characteristic of the change that took place from the 1990s to the 2000s was that the percentage of environmental conservation expenses allocated to the Ministry of Construction and the Ministry of Transport (now Ministry of Land, Infrastructure, Transport, and Tourism) fell to 48 percent, while that going to the Ministry of Agriculture, Forestry and Fisheries and the Environment Agency (now the MOE) increased.

Table 3.5 Ministries by proportion of the environmental conservation budget (central government) (unit: %, JPY 100 million)

Ministry[a]	Responsibilities (before 2011)	1972	1975	1980	1985	1990	1992	1995	2000	2005	2009	2010
Cabinet Office (CAO)	Atomic Energy Commission; Nuclear Safety Commission; Council for Science, Technology and Innovation.	1.0	1.4	1.4	1.4	1.2	1.5	1.5	4.3	5.3	2.4	2.3
Ministry of Foreign Affairs (MOFA)	Diplomatic policy and negotiations relating to global environmental issues.	0.0	0.0	0.0	0.0	0.0	0.0	0.1	0.2	0.2	0.3	0.5
Ministry of Education, Culture, Sports, Science and Technology (MEXT)	Environmental education; protection of cultural heritage and natural environment.	2.7	3.4	1.5	1.3	1.1	1.0	10.8	8.9	6.5	4.0	5.6
Ministry of Health, Labour and Welfare (MHLW)	Public health, life health, water supply administration	5.0	6.5	5.8	5.7	4.8	6.5	5.3	0.2	0.2	0.1	0.3
Ministry of Agriculture, Forestry and Fisheries (MAFF)	Management of natural forests, conservation of fishery resources, promotion of sustainable agriculture, regulation of agricultural chemicals.	3.1	3.6	2.3	2.0	2.1	7.3	9.6	17.1	15.8	17.0	18.9

Ministry of Economy, Trade and Industry (METI)	Promotion of energy conservation, development of technology for industrial pollution prevention and control, recycling of industrial waste.	2.5	5.5	2.5	1.1	0.7	0.6	0.6	0.8	10.8	16.1	27.2
Ministry of Land, Infrastructure and Transport (MLIT)	Control of pollution form road vehicles, development of public works (e.g. sewerage, urban parks, and roads), restoration of rivers, prevention of coastal zone pollution.	71.4	65.5	75.1	75.3	59.2	71.0	53.3	48.2	51.2	46.3	22.8
Ministry of the Environment (MOE)	General environmental conservation	4.7	6.1	3.8	3.8	3.7	3.7	2.7	8.5	9.9	10.5	16.7
Other ministries		9.6	8.1	7.7	9.3	27.3	8.4	16.0	11.9	0.0	3.2	5.8
Total (%)		100.0	100.0	100.0	100.0	100.0	100.0	100.0	100.0	100.0	100.0	100.0
Total (JPY 100 million)		1,693	3,751	11,663	11,172	13,402	15,514	25,986	30,419	23,604	21,168	12,595

a Reclassified on the basis of the organization of ministries after the reorganization of 2001.

Source: Author's calculations based on data from the Ministry of Environment.

The Basic Environment Law was enacted in 1993 in response to global environmental problems, destruction of nature, and lifestyle and urban pollution. However, there was no statement of change with regards to the business of the Environment Agency. The Environment Agency's duty and power were enhanced and strengthened when it was reorganized into the Ministry of the Environment (MOE) as part of the restructuring of Japan's central government on 6 January 2001.

The MOE was tasked with pollution prevention, maintenance of the natural environment, and other environmental conservation tasks. In addition to these traditional tasks, conservation of the global environment now fell under its jurisdiction according to a written statement. Moreover, the administration of waste matter processing was transferred to the MOE from the Ministry of Health and Welfare (now the Ministry of Health, Labour and Welfare). The MOE also strengthened its power and mission by centralizing some types of conservation that it had previously shared with the Ministry of Agriculture, Forestry and Fisheries (MAFF).[15]

After 2000, the amount of the environmental conservation budget controlled by the MOE increased from its previous 3 to 5 percent to 9 to 17 percent. However, this is not a high percentage, and many different ministries and government offices are still involved in environmental policy. In other words, when environmental conservation budget requests are made, the MOE adjusts the budget requests of the other ministries and government offices. The MOE is still required to give adequate consideration to the environmental conservation policy of other ministries and government offices.

For example, the allocation of eco-points for consumer electronics (see Chapters 4 through 7) is under the jurisdiction of the MOE, the Ministry of Economy, Trade and Industry (METI), and the Ministry of Internal Affairs and Communications (MIAC).[16] Moreover, the 2009 eco-car subsidies (see Chapters 8 and 9) are under the jurisdiction of the METI and the Ministry of Land, Infrastructure, Transport and Tourism (MLIT). They are not under the jurisdiction of the MOE.[17] In other words, these environmental policies are not decided independently by the MOE.[18]

The basic environmental plan that is developed by the MOE is approved by the cabinet, and it becomes the guideline for the central government's budgetary allocations for environmental policy.[19] However, the plans of each ministry, their relationship to the basic environmental plan, and the priority of different programs are not necessarily clear. Moreover, the environmental policy of local governments is not controlled by the central government. In short, despite the fact that the basic environmental plan is approved at the cabinet level, it is currently difficult to make the claim that the basic environmental plan has become a consistent framework for the activities of local governments and all of the ministries and government offices (OECD 2010, 24).

3.3.1.2. *The current state of environmental expenditure policy*

Table 3.6 shows the percentage of expenditures of the central government's environmental conservation budget controlled by different ministries and government

Table 3.6 Environmental conservation budget (central government): percentage of different ministries in each area in 2009 (unit: %, JPY 100 million)

Ministry	Realization to the recycling economy and society						Ensure coexistence of man and nature	Others	Total amounts (unit: JPY 100 million)
	Conservation of the global environment	Conservation of the atmospheric environment	Conservation of water environment, soil environment, and ground environment	Waste and recycling measures	Chemical substance control				
Cabinet Office (CAO)	2.3	0.1	9.6	1.8	2.8	–	3.8	0.2	504
Ministry of Foreign Affairs (MOFA)	0.3	0.8	–	–	–	–	–	–	56
Ministry of Education, Culture, Sports, Science and Technology (MEXT)	3.8	8.7	–	1.2	–	–	6.3	0.7	849
Ministry of Health, Labour and Welfare (MHLW)	0.2	0.1	–	0.0	–	22.8	–	–	31
Ministry of Agriculture, Forestry and Fisheries (MAFF)	13.3	22.8	–	8.8	12.9	19.7	42.4	16.5	3,604
Ministry of Economy, Trade and Industry (METI)	19.1	48.8	1.5	0.4	1.4	12.7	0.0	2.5	3,417

(Continued)

Table 3.6 (Continued)

Ministry	Realization to the recycling economy and society							Others	Total amounts (unit: JPY 100 million)
		Conservation of the global environment	Conservation of the atmospheric environment	Conservation of water environment, soil environment, and ground environment	Waste and recycling measures	Chemical substance control	Ensure coexistence of man and nature		
Ministry of Land, Infrastructure and Transport (MLIT)	49.1	10.6	59.9	87.4	9.7	0.1	41.2	0.3	9,803
Ministry of the Environment (MOE)	8.1	7.9	1.0	0.3	72.3	44.6	6.2	79.1	2,218
Other ministries	3.8	0.1	28.0	0.1	0.8	0.0	0.1	0.7	687
Total (%)	100.0	100.0	100.0	100.0	100.0	100.0	100.0	100.0	
Total amounts (JPY 100 million)	17,777	6,780	2,342	7,432	1,140	82	2,612	780	21,168
Share (%)	84.0	32.0	11.1	35.1	5.4	0.4	12.3	3.7	100.0

Source: Author's calculations based on data from the Ministry of Environment.

offices in the various environmental policy areas in 2009. In 2009, "global environment conservation" comprised 32 percent of environmental conservation expenditures. This was 48 percent of the METI's budget and 22 percent of the MAFF's budget. "Conservation of water, soil, and land environments" comprised 35 percent of environmental conservation expenditures. This was 87.83 percent of the MLIT's budget. Also, "natural environment conservation" comprised 12 percent of environmental conservation expenditures. This was 42 percent of the MAFF's budget and 41 percent of the MLIT's budget.

It has been previously pointed out that since 2000 there has been an overall increase in funding for "global environmental conservation" and a decrease in that for "conservation of water, soil, and land environments." Also, when this is viewed by ministry and government office, it becomes apparent that there has been an increase in funding for "global environmental conservation" in the METI's percentage and a decrease in that for "conservation of water, soil, and land environments" in the MLIT's percentage.[20]

The scope of the MOE's budget has gradually expanded, and it is now 72.3 percent of the amount allocated for "waste matter and recycling measures" and 44.6 percent of that allocated for "measures for chemical substances." But these fields are relatively low percentage among environmental conservation expenses. In other words, the MOE budget is guaranteed by absorbing tasks and programs related to the environment that have been discarded by other ministries and government offices or by taking on new regions that other ministries and government offices have not traditionally handled.

3.3.2. Environmentally related taxes in government departments

3.3.2.1. The formation of environmental taxation policy and the Ministry of the Environment

The MOE must coordinate with other ministries and government offices when debating environmentally related taxes. In general, each ministry works out the plan for requests for tax revisions of national taxes. The process is such that each ministry submits its requests to the Ministry of Finance. Moreover, each ministry works out a plan for requests for tax revisions of local taxes. These requests are then submitted to the Ministry of Internal Affairs and Communications (MIAC).[21] However, adjustment of requests for revision of the environmentally related taxes by the MOE is much more complicated due to adjustments to expenditures. This is because the environmentally related taxes are connected to a system of specified financial resources. A bit more detailed explanation of this is that ministries and government offices that retain specific financial resources (in this case the Ministry of Land and Ministry of Economy) are required to apply funds based on certain taxation laws or by some other special law. In this way they acquire spending power, and through that spending power they have influence over annual revenue (taxation system revision).

Japan's taxes are shifting toward environmentally related taxes. For example, Japan's vehicle and energy taxation systems traditionally did not take the environment into account, but today they are now in the process of incorporating environmentally related taxes. An understanding of this change can be obtained by examining the position of the MOE as a coordinating ministry.

One primary reason for the tipping point in the change of the MOE's position in recent years has been the debate on the revision of the specified financial resource system. The policy of reviewing the system of specified financial resources starting with financial resources that were exclusively set aside for road construction was started in April 2001 with the creation of the Koizumi administration. The reason that the specified financial resource system became the subject of debate was that, in the midst of an expanding budget deficit, the nation viewed the specified financial resource system not having enough for other expenditures as a problem.

3.3.2.1.1 THE HISTORY OF MAKING VEHICLE TAXATION GREEN

The history of making vehicle taxation green stems from the cancellation of limitations on specified financial resources. Due to this, participation in the MOE's tax reforms became easy.

The first phase was making the automobile tax (a general financial resource that was a local tax) that was implemented in 2001 green while maintaining the system of specified financial resources.[22] This was the first tax reform that consciously included environmental considerations within the taxation system.

For the purpose of this work, more focus should be given to the lessons learned from the failed introduction of tax reform in 1999 and to two amendments made to the taxation system in 2000.[23]

The first amendment to the tax system was removing the automobile weight tax as a target of tax reform investigations. In the requests for tax reform in 1999, the automobile weight tax was subject to examination, but the Ministry of Construction (currently the MLIT) was opposed to reforms of the automobile weight tax that would require funds raised by the tax be used to pay specifically for highway construction. The removal of the automobile weight tax from the request for tax reform in 2000 showed that it would be difficult to start work on taxes that would become specified financial resources. Thus, it could be said that only the automobile tax – a local tax and a general financial resource – was the subject of investigation.

Another amendment was a joint plan by the Ministry of Transport (currently the MLIT), the Environment Agency, and the Ministry of International Trade and Industry (currently the METI) that addressed both fuel economy standards and exhaust fume emission standards.[24] Both the Environment Agency and the ministry made demands for tax reform in 1999, but the Environment Agency was not satisfied that such reforms did not consider exhaust fume emission performance.[25] Also, the Ministry of International Trade and Industry was concerned that stronger taxation on high-profit-margin luxury cars due to fuel economy standards would have a negative impact on automotive sales.[26] This dissatisfaction and worry was alleviated due to the addition of exhaust fume emission performance standards in the tax reform of 2000. It could be said that the Environment Agency's demands for environmental considerations were accepted.

The second phase was the decisions in the debate on 2009 tax reform that took place starting in the fall of 2008. The so called eco-car tax reduction was determined and automobile weight tax and the automobile acquisition taxes became environmentally friendly. Because the details and effects of this system are analyzed in Chapter 9 of this text, only these characteristics will be pointed out here.[27]

Several background factors that set the scene for these reforms was the cabinet decision made in May 2008 that financial resources that were exclusively set aside for road construction would become general financial resources starting in 2009 and the global economic crisis that was precipitated by the failure of Lehman Brothers in September of 2008 that spurred the demand for economic measures.

The automobile weight and automobile acquisition taxes that had been financial resources that were exclusively set aside for road construction (and as such had been difficult to reform in the past) were reduced. Also, up to this point, the taxation system had only given environmental considerations to local taxes, but now it also gave environmental considerations to national taxes. These reforms were more of a political decision than they were product of debate within the administrative organizations. However, this suggests that the participation of the MOE in the environmentally related tax reform process became much easier when compared with the past. This was due to the shifting of financial resources that had been exclusively set aside for road construction to general financial resources. The shifting of financial resources that were exclusively set aside for road construction into general financial resources was a turning point for environmental considerations being incorporated into vehicle taxes.

The third phase was decisions that emerged from the debate of 2014 tax reform that took place starting in the fall of 2013. In this debate, it was decided that vehicle taxes would be made much more green (i.e. incorporate more environmental considerations). Specifically, it was determined that the automobile weight tax, a national tax that was formerly a specified financial resource, would be made much more green and that the automobile acquisition tax, a local tax and formerly a specified financial resource, would be abolished.

In the background of these decisions regarding reforms, there was the demand for simplification of the tax code, a reduction in the tax burden, and change to more environmentally friendly taxes when the consumption tax rate increased from 5 to 10 percent.

The characteristics of these reforms on the taxation system were that both national taxes and local taxes that were formerly collected specifically to pay for highway construction were reformed. To avoid impacting the financial resources of local governments, the local automobile acquisition tax was formally abolished, which increased the possibility of it being integrated into a much more environmentally friendly local automobile tax.[28]

During this tax reform process, the MOE clearly expressed its stance of wanting vehicle taxes to be made much greener, but it did not make any concrete demands. In contrast, the METI and the MLIT made specific requests for almost the same content. They asked for an expansion of the reduction for the automobile weight tax (a national tax),[29] a graded reduction in the automobile acquisition tax (a local tax), an abolition of the automobile acquisition tax when the consumption tax is at 10 percent,[30] and the expansion and extension of the reduction for the automobile tax (a local tax). This suggests that even if the MOE does not actively pursue specific demands, it has become easier to make progress on making things greener.

3.3.2.1.2 THE HISTORY OF MAKING ENERGY TAXATION GREEN

In the history of making energy taxation more environmentally friendly (green), the MOE expanded its participatory power in the tax reform process by entering into the specified financial resource framework of special account for energy. This was established by the METI during a period rapid economic growth.

The first phase of this history started in the debate on tax reform for 2003 that began in the fall of 2002. In this debate it was determined that there would be new taxes on petroleum that would start on 1 October 2003. The traditional petroleum tax (a national tax) was changed to the petroleum and coal tax (also a national tax).

Under the Koizumi administration that began in April 2001, more opportunities to review the special account came about because of a surplus of financial resources caused by the fact that these taxes were specified financial resources. In the midst of this, there was pressure to reform the special energy account that was under the jurisdiction of the METI. The METI found that it was essential for there to be a reduction in surplus funds via an expansion of the purposes for which money was spent in order for there to be the continued existence of a special energy account. The point of interest here was the conversion of the structure of annual expenditure to one that took environmental considerations into account, as advocated by the MOE. By expanding money spent on energy policies useful for controlling the carbon dioxide emissions that accompany energy usage, the METI shared control of the special energy account with the MOE. At the same time, the METI tried to levy new taxes on coal. The logic for new taxes on coal was such that if there was to be a review of the annual expenditure structure (benefits) of the special account system that as a general rule burdens beneficiaries, it was essential to simultaneously review the annual income structure (burden). This was done so that the benefits and burdens of the consumers of every type of fuel would not be separated.

The MOE had the opinion that this call by the METI was a clever scheme by the METI to destroy the green tax plan that had been investigated by the MOE primarily as a means to control carbon dioxide emissions.[31] After much discussion, the MOE and METI cabinet ministers came to an agreement that introducing a petroleum and coal tax by reforming the special energy account and the implementation of a green tax that was primarily motivated to control carbon dioxide emissions were not related.[32]

The change from the traditional petroleum tax to a petroleum and oil tax (a national tax) was a way for the METI to reform the special energy account by involving the MOE. Moreover, it was a way for the METI to hold onto energy policy leadership by creating a "real" green tax before the MOE could introduce its own green tax.[33] The MOE was able to expand its expenditure authority for specified financial resources related to the environment and its participatory power on the petroleum and coal tax that is connected to the environment. This was done when the MOE gained joint control of the special energy account that was under the METI's jurisdiction. At that time, the agreement that the green tax was a separate matter from the petroleum and oil tax was very important to the MOE.

The second phase of this history was the submission of a green tax plan in the fall of 2004 (a national tax) that had the primary purpose of controlling carbon dioxide emissions in the tax reform requests for 2005. The MOE had previously deliberated on green tax plans, but there had yet to be a concrete submission to the Ministry of Finance and MIAC during the process of taxation system reform. In the tax system reform requests for 2005 that took place in the fall of 2004, the MOE made its first submission of the green tax plan to the Ministry of Finance in the form of a joint proposal with the MAFF.

The features of the green tax plan submitted in November of 2004 were that its primary objective was to control carbon dioxide emissions; it was not an adjustment of the existing energy taxation system but was rather new taxation; and it allocated the majority of its tax revenue to global warming mitigation measures. The METI was opposed to this because it wanted to maintain its leadership on energy policy and was worried that "it would lose the trust built with the industrial world if a tax hike was approved again so soon after the tax increase caused by the petroleum and coal tax."[34] The MLIT did not take a position of clear opposition. This was because the plan was not a change to the existing energy taxation system and it did not impact the taxes collected specifically to pay for highway construction. However, in the end, the MOE's green tax plan did not have strong backing from the supporting parent organizations, so its introduction fell to the wayside.[35]

The third phase was the introduction in October of 2012 of a tax that would fund global warming countermeasures (a national tax).[36] Discussion of this tax policy spanned the tax reform debates for 2010 and 2011 that took place in the fall of 2009 and 2010, respectively.

A number of factors were in the background of the two-year debate on the tax for global warming countermeasures. First, in May 2008, a cabinet decision was made that financial resources that were exclusively set aside for road construction would become general financial resources beginning in 2009. Second, in the fall of 2009 a Democratic Party administration that placed importance on environmental policy came into power.

The Democratic Party's intentions were reflected in the fall 2009 requests for 2010 tax system reform. In response to the MOE presenting a green tax plan, the METI also presented a green tax plan. Both of these green tax plans shared the point that taxes would be levied and collected based on fossil fuel use in the form of extra taxes for the petroleum and coal tax. The difference in the two ministries' plans was that in the MOE's plan the tax revenue did not specify the purpose for which money would be spent, whereas the METI's did.[37]

The MOE claimed that the introduction of a green tax was necessary to achieve a 25 percent reduction in carbon dioxide emissions. However, it was also aware of the sequence of events that led to the financial resources that were exclusively set aside for road construction becoming general financial resources. The MOE considered carefully whether to make the green tax a specified financial resource.[38] In the end, the purpose for which the tax revenue would be spent was left unspecified, and it became a general financial resource. However, the MOE stated that the annual expenditure and tax reduction for global warming countermeasures that had an effect on reducing carbon dioxide emissions should be given priority. In contrast, although the METI claimed that the introduction of a green tax would be necessary to achieve a 25 percent reduction in carbon dioxide emissions, such a tax should be thought of as just part of a larger energy policy. Such a tax required a thorough understanding of the industrial sector and the nation as whole and thus the purpose for which money would be spent should be specified. In METI's view, if the MOE turned the green tax revenue into a general financial resource, then it should be done completely separately from the petroleum and coal tax (a specified financial resource). In other words, the METI bluntly refused the MOE's plan.[39]

In the requests for reform of the 2010 taxation system, the decision on the tax for global warming countermeasures was deferred. The MOE then amended its plan and started work amidst the request for reform of the 2011 taxation system that took place in the fall of 2010. The MOE changed its plan so that the extra tax revenue levied and collected through the petroleum and coal tax would not be a general financial resource, but rather would be allocated as a financial resource with a specific purpose.[40] By not specifying the purpose for which the tax revenue was spent and turning it into a general financial resource in the tax reform process of the previous year, the MOE tried to gain support with other government ministries and offices.[41] However, it then decided that this was the reason it had been unable to get support from these other ministries.[42]

It was through this process that the tax for global warming countermeasures was introduced in steps in the form of the surplus of the petroleum and coal tax (national tax) that was levied after October 2012. The tax for global warming countermeasures was introduced in a form that actualized the shared portions of the plans of the MOE and the METI. The tax revenue that it generated was a financial resource that was specified for the special energy account.

The METI retained a special energy committee through the introduction of the tax for global warming countermeasures while acquiring additional specified financial resources for energy measures. The MOE acquired the power to participate in the tax for global warming countermeasures through the special energy committee's specified financial resource by having the METI's claim pushed on them.[43] However, when the petroleum and coal tax was created, there was an agreement made in name but not in reality in November 2002 between the MOE and METI. This agreement came in the form of a compromise by the MOE which stated that the petroleum and coal tax was separate from the green tax that had been investigated by the MOE.

3.3.2.2. *The current state of environmental taxation policy*

As we have already confirmed, environmentally motivated taxation policy after 2000 saw a shift within the taxation system toward connecting environmental policy with national and local taxes. This was a shift from complete devotion to reducing the tax burden to revising the existing taxation system and introducing new taxes. In particular, the vehicle and energy taxes that had given little consideration to the environment were changed to repressive taxes that sought to reduce environmental burdens. This was largely due to the debate on revising the system of specified financial resources and the change of government.

Finally, we will gain an understanding of the features of Japan's current environmental taxation policy by monetarily verifying the available range of special taxation measures related to the environment that are defined differently from the OECD's definition of an environmentally related tax[44] (Table 3.7).

Table 3.7 Current status of environment tax policy by special taxation measures related to the environment[a]

	Level of Tax	*(Unit: JPY 100 million[b])*	
		2011	*2012*
Realization of low-pollution type society		▲ 97.5	▲ 83.0
Reduce the corporate income tax on pollution-control equipment. (Special depreciation)	National	▲ 2.4	▲ 0.0
Reduce the local corporate income tax on pollution-control equipment. (Special depreciation)	Local[c]	▲ 1.0	▲ 0.0
Reduce the business facility tax on pollution-control equipment. (Special provisions for tax base)	Local	▲ 3.6	▲ 3.4
Reduce the real property tax on pollution-control equipment. (Special provisions for tax base)	Local	▲ 90.5	▲ 79.6
Environment-related investment promotion		▲ 481.5	▲ 663.9
Reduce the corporate income tax for energy supply and demand structure reform. (Special depreciation)	National	▲ 338.4	▲ 131.3
Reduce the local corporate income tax for energy supply and demand structure reform. (Special depreciation)	Local[c]	▲ 132.9	▲ 60.1

(Continued)

Table 3.7 (Continued)

	Level of Tax	(Unit: JPY 100 million[b])	
		2011	*2012*
Reduce the corporate income tax for energy supply and demand structure reform. (Special deduction)	National	▲ 5.0	▲ 2.0
Reduce the local corporate income tax for energy supply and demand structure reform. (Special deduction)	Local[c]	▲ 0.9	▲ 0.4
Reduce the corporate income tax on investment related to the environment. (Special depreciation)	National	▲ 2.4	▲ 325.4
Reduce the local corporate income tax on investment related to the environment. (Special depreciation)	Local[c]	▲ 0.7	▲ 137.6
Reduce the corporate income tax on investment related to the environment. (Special deduction)	National	▲ 1.0	▲ 6.0
Reduce the local corporate income tax on investment related to the environment. (Special deduction)	Local[c]	▲ 0.1	▲ 1.1
Measures against global warming		–	924.0
Special provisions for tax rate of petroleum and coal tax. (Addition of tax rates according to CO_2 emissions)	National	–	924.0
Greening of the vehicle body taxation		▲ 3,440.7	▲ 2,772.5
Review of the automobile weight tax on motor vehicles with excellent environmental performance. (Eco-car tax reduction)	National	▲ 1,018.0	▲ 347.0
Reduction of the automobile acquisition tax on automobile with excellent environmental performance. (Eco-car tax reduction)	Local	▲ 2,347.2	▲ 2,329.5
Special provisions for greening of the automobile tax on motor vehicles with excellent environmental performance. (Heavy taxation)	Local	237.3	250.0
Special provisions for greening of the automobile tax on motor vehicles with excellent environmental performance. (Light taxation)	Local	▲ 312.7	▲ 346.0
Appropriate disposal of waste		▲ 9.2	▲ 8.6
Exempt/reduce the business facility tax for waste processing facilities. (Special provisions for tax base)	Local	▲ 9.2	▲ 8.6

Promotion of renewable energy		–	0.0
Reduce the real property tax on renewable energy power generation equipment (Special provisions for tax base)	Local	–	0.0
Biofuel		▲ 195.0	▲ 188.0
Special provisions concerning gasoline tax to ethanol-blended gasoline. (Deduction of the part that is mixing)	National	▲ 195.0	▲ 188.0
Energy-saving housing		▲ 6.4	▲ 5.8
Reduce the individual income tax on investment related to energy-saving housing. (Special tax credit)	National	▲ 5.0	▲ 5.0
Reduce the real property tax on housing saving renovation.	Local	▲ 1.4	▲ 0.8
Total Amounts (JPY 100 million)		▲ 4,230.3	▲ 2,797.8
Tax cut from normal national tax related to the environment		▲ 1,567.2	▲ 80.7
Tax cut from normal local tax related to the environment		▲ 2,663.1	▲ 2,717.1

a There are tax cut measures and tax increase measures in the special taxation measures related to the environment. The ▲ means the amount of tax cut.
b Local tax is actual value. Due to limitations in the data, the actual value of the national tax is only corporate income taxes. The other national tax are estimated at the time of budgeting.
c Contains a special local corporate tax, which influences the amount of local corporate income tax.

Source: Author's calculation based on data from MIAC (2014), Ministry of Finance (2014), and METI (2011, 2012).

First, as of 2011, tax breaks that were representative of the era of industrial pollution that went to pollution-preventing institutions totaled a postponement 240 million yen of national corporation taxes, reducing local fixed asset taxes by 9.05 billion yen. In contrast, the estimated amount of money that was a postponement of corporation tax from the era of industrial pollution was 33.8 billion yen in 1972 and 61.0 billion yen in 1975[45]; therefore, there has been a striking decrease in tax breaks for pollution-preventing facilities.

The taxation system for promoting investment in the reform of the energy supply and demand structure postponed 33.84 billion yen of national corporation taxes and 13.29 billion yen in local corporation taxes in 2011. This was a total of 47.13 billion yen. In other words, there was a shift in promotion of investments for the environment from pollution-preventing facilities to the energy-efficient equipment and alternative energy equipment.

Second, the amount of money for the 2011 national tax special measure items was 156.72 billion yen, and the amount of money for the local tax special measure items was 266.31 billion yen. The amount of money for special measures that used local taxes was larger. Almost all of the local tax reductions were done within the uniform countrywide framework.

Third, the increase in revenue due to heavy automobile tax measures was 23.73 billion yen, and the increase in revenue due to the introduction of taxes for global warming countermeasures after October 2012 was 92.4 billion yen. Thus, there was a shift from a complete devotion to decreasing the tax burden to passing repressive taxes that were more environmentally aware (i.e. that internalized external diseconomies).

Fourth, the largest amount of money within special measures related to the environment was the special measure for vehicle taxation. The amount of tax reduction in 2011 was 344.0 billion yen for the combined total of national and local taxes. According to currently available data, this comprises approximately 81 percent of the 423.0 billion yen used for special measures related to the environment.

The so-called eco-car tax reduction in 2011 was 101.8 billion yen in the form of the national automobile weight tax and 234.7 billion yen in the form the local tax automobile acquisition tax. This is a total amount of 336.5 billion yen. According to currently available data, this comprises approximately 79 percent of the total amount of 423.0 billion yen used for special measures related to the environment.

Finally, manufacturers have traditionally been able to take advantage of a number of special measures, such as reductions in business facility taxes and fixed asset business taxes. In recent years, special taxation measures (tax cuts as well as tax increases) related the environment have been targeted at manufacturers as well as at consumers, such as recent vehicle taxation laws. These measures have sought to address the striking increase in lifestyle and urban pollution (household pollution).

3.4. Conclusion

This chapter described the monetary basis of taxation and the annual expenditure policies with regards to Japan's environmental policy. It also explored the process by which Japan's environmental policy is made. This chapter will be concluded by a brief summary that parallels the historical changes in environmental policy.

Industrial pollution from factories had become a problem. The primary environmental policy used to address industrial pollution was to individually enumerate and legally regulate the actions and facilities that directly impacted health due to the environmental burden they caused. Services and facilities that contributed to environmental conservation but were not adequately provided by the private sector were directly provided by the local and central governments. Moreover, businesses that faced these legal regulations had their tax burden reduced by subsidies, tax-reduction measures, and favorable interest rates via the Fiscal Investment and Loan Program. These strategies were implemented by the government in order to support private-sector investments in pollution control.

Environmental expenditure policy has evolved over time through the establishment of the Environment Agency and the implementation of various anti-pollution measures. The central government's budget for environmental conservation expenses has changed over time. At the beginning of the 1970s, local governments had more money for anti-pollution measures than the central government had for environmental conservation expenses. This was because the local governments played a large role in the initial legal restrictions to eliminate pollution. Many of the local governments had established stricter emissions standards than those created by the central government.

Tax reductions were carried out as part of revisions to taxation policy. The target of these tax reductions was primarily the companies that were involved in manufacturing. A reduction (or postponement) of the corporation tax (a national tax) and the fixed assets tax (a local tax) was implemented to reduce the burden on companies that invested in anti-pollution measures. Awareness so-called environmentally related taxes (taxes that give consideration to the environment) was exceedingly low at that time.

Measures to address household pollution resulting from air pollution due to automobile emissions and waste management required different environmental policies. Historically, these techniques had generally consisted of the central government and local governments implementing environmental conservation measures while offering assistance to the businesses (manufacturers) that were regulated. New measures sought to address consumer behavior and purchases.

In the beginning, local governments were the principal players in implementing policies with environmentally conscious expenditures. However, the majority of financial resources for environmental measures by local governments were subsidies provided to the local governments by the central government. Thus, when the amount of money or the standard for subsidies related to environmental conservation for localities by the central government changed, the amount of money for environmental conservation expenditure by both the local and central governments were impacted.

Moreover, by connecting environmental policy with national and local taxes, the taxation system shifted from a complete devotion to reducing the tax burden for those who were legally regulated to the renovation of the existing taxation system and introduction of new taxes. In particular, after 2000, vehicle and energy taxes that had been developed with little consideration of the environment became repressive taxes that sought to control environmental burdens (i.e. internalizing external diseconomies).

The MOE is the central government ministry that is tasked with developing and carrying out a comprehensive environmental policy. However, it has been merely given the power to comprehensively coordinate environmental policies and actions put forth by other administrative organs. The MOE requests that other ministries and government offices give adequate consideration to both environmentally motivated expenditure policy and taxation policy. However, with regard to taxation policy, it has become easier than in the past for the MOE to voice its opinion in the tax reform process. This is because of the influence of the debate on the use of specified financial resources.

Acknowledgment

The author received assistance for developing this chapter from JSPS Research Funds/Grant-in-Aid for Challenging Exploratory Research "International comparative analysis related to budget and auditing systems as well as the financial resource relationships between governments and the public financial system" (research representative: Satoshi Sekiguchi issue number: 24653106).

Notes

1 Abe and Awaji (2006), p. 61, Ueta (2002).
2 In addition to the domestic environmental policy dispute, there are also other discussions, such as the finance policy of the Fiscal Investment and Loan Program and the international environmental policy of the Official Development Assistance (ODA) expenditure (Lee 2004). However, these are outside the scope of this chapter.
3 The OECD defines "environmentally related taxes" as all forced and one-sided payments to a government that are levied on things considered to be related to the environment in particular (OECD 2001). Thus, the purpose and name of the tax is not the criteria nor is the decision on the purpose for which the tax money is spent. The targets of taxation for these environmentally related taxes are things such as "energy products," "automobiles and other means of transportation," "waste matter processing," and "substances that destroy the ozone layer" (OECD 1977; Miyajima 1993; Morotomi 2002b; Yoshimura 2008; Fujitani 2010; Jinno 2013; Mochida 2013).
4 Ministry of Environment (2002).
5 Calculations taken from each year's "Public Finance Monthly Statistical Report (Budget Report)".
6 The decrease in the percentage of the environmental conservation expenses by the Ministry of Land after 2010 was influenced by the generalization of the funding source for the revenue set aside for road construction. In order to supplement the portion of expenditure used on localities by the central government's social capital maintenance special account (which accompanied the change in the funding source for the revenue set aside for road construction) the "comprehensive subsidy for the maintenance of social capital" was established within the central government's general accounting and finance in 2010. This subsidy was excluded from environmental conservation expenses because it could not be specified as an environmental activity by the local governments.
7 The expenses by local governments for anti-pollution measures do not include the expenses for measures taken to combat global warming (Ogata 2006). This is because under the Basic Environment Law they are categorized not as "pollution" expenses but as "global environmental conservation" expenses. The initial budget for 2008 had the amount of money spent by prefecture governments at 785.2 billion yen and the amount of money spent by city governments at 720.7 billion yen for a sum total of 1.5058 trillion yen (per Select Committee on the Taxation System material that was submitted to the Ministry of Internal Affairs and Communications, 18 November 2009) (Hayashi 2004).
8 Calculation from the 1996 *Annual Report on the Environment*, article 6, section 3.
9 Calculation from the 1998 *Annual Report on the Environment*, article 6, section 3.
10 The timing was slightly after the revision of the tax system, but in the revised budget of April 2009 the eco-car subsidy was introduced. This meant that there were cases where the eco-car tax reduction and the eco-car subsidy could be used together.
11 The tax on industrial waste is an earmarked tax that is outside the stipulations of local tax law. The green forest tax is in practice an earmarked tax because the usage of the tax revenue it generates is limited. It is a tax that goes beyond ordinary private municipal taxes (Morotomi 2002a; Takai 2013; Yokoyama 2014).
12 The tax revenue generated by the tax on industrial waste in 2012 was 6.9 billion yen. The tax revenue generated by the green forest tax was approximately 23 to 24 billion yen.

13 Abe and Awaji (2006), pp. 10–11, 14.
14 Departments that were closely related to the environment were integrated, such as the headquarters on anti-pollution measures, the Ministry of Health and Welfare, the Ministry of International Trade and Industry, the Economic Planning Agency, and the Forestry Agency.
15 Abe and Awaji (2006), p. 15.
16 The grand total of the budget for the eco-points program was approximately 693 billion yen. In the first revised budget of 2009, it was approximately 294.6 billion yen. In the second revised budget of 2009, it was approximately 232.1 billion yen. It was approximately 85.5 billion yen in the 2010 emergency funds for regional revitalization/coping with the economic crisis, and it was 77.7 billion yen in the revised budget of 2010.
17 The grand total of the budget for eco-car subsidies was approximately 580 billion yen. It was approximately 357.2 billion yen in the first revised budget of 2009, and it was approximately 230.4 billion yen in the second revised budget of 2009. Moreover, new eco-car subsidies are at 300 billion yen under the fourth revised budget of 2011.
18 Unfortunately, the amount of money for the eco-points program and eco-car subsidies were not in the initial budgets, so they are not included in the environmental conservation expenses shown here.
19 Moreover, the basic environmental plan is being implemented through regional plans and plans made for each field.
20 Reference footnote 6.
21 The debate on tax reform for Japan's next fiscal year start in the fall of the present year, and usually the cabinet decision is made around January of the next year.
22 The Tokyo Metropolitan area made the decision on making the automobile tax green in March 1999 via excessive and non-uniform taxation. It was independently implemented in 2000, and it was introduced nationwide in 2001.
23 For other policy content, please reference Ito (2010), pp. 137–40.
24 In cases of light-class vehicles. For the heavier class, fuel economy standards change several years after a new car is registered.
25 MOE demanded an independent plan that gave a standard for exhaust fume performance as an anti-air pollution measure.
26 Ito (2010), p. 138.
27 Reference footnote 10.
28 The local vehicle taxes were moving in a direction where they could be organized into the automobile taxes (prefecture taxes) that were originally a general financial resource and that had the characteristics of property taxes and light-vehicle taxes (city taxes).
29 MLIT asserts that the basic structure for the eco-car tax reduction via the automobile weight tax (a national tax) is permanent.
30 METI asserts that the consumption tax for the vehicle acquisition tax (a local tax) will be lowered 3 percent when it is at 8 percent and when the consumption tax is at 10 percent it will be abolished.
31 Aoki and Suzuki (2007), p. 130.
32 Please reference the MOE's homepage (http://www.env.go.jp/policy/info/energy/index.html) for the text of this agreement.
33 Aoki and Suzuki (2007), p. 129.
34 Aoki and Suzuki (2007), p. 130.
35 During this period, politicians began to target environmental taxes for investigation (specifically by the LDP's Selective Committee on the Taxation System). There were also investigations into limiting the use of tax revenue to measures taken for global warming. Moreover, MOE made a proposal to present for the

2006 tax system reform requests in the fall of 2005 an adjustment of the existing energy taxation system, but its introduction also fell to the wayside.

36 This tax as a global warming countermeasure was presented to the National Diet as a legal plan based on the fundamental principles of the reform of the 2011 tax system decided by the cabinet on 16 December 2010. However, there was also the influence of the Great Tohoku Earthquake that happened on 11 March 2011, and the result of the National Diet's deliberations was that the items for reform fell to the wayside. A bill was passed in March 2012 as a reform of the 2012 taxation system.

37 MOE's plan clearly positioned the tax revenue that corresponded to the temporary tax rate on gasoline as a measure taken for the purpose of controlling carbon dioxide emissions. When paying attention to the items of taxation that were formerly financial resources that were exclusively set aside for road construction, the MOE's plan references only the deferment of gasoline taxes and the temporary tax rates for local gasoline taxes. It does not cover light oil delivery taxes or vehicle taxation (the automobile weight tax and vehicle acquisition tax). METI's plan, however, did not deal with the treatment of items of taxation that were formerly financial resources that were exclusively set aside for road construction.

38 Select Committee on the Taxation System Minutes from the 20th meeting (7 December 2009).

39 Select Committee on the Taxation System Minutes from the 20th meeting (7 December 2009).

40 The tax revenue from the temporary gasoline tax rate was clearly for the purpose of controlling carbon dioxide emissions.

41 Select Committee on the Taxation System Minutes from the 6th meeting (5 November 2009).

42 Select Committee on the Taxation System Minutes from the 6th meeting (2 November 2010).

43 The ministries and government offices connected to the environmental policy of the Ministry of Agriculture Forestry and Fisheries are still requesting an expansion of the purpose for which this money is spent.

44 There are tax cut measures and tax increase measures in the special taxation measures related to the environment. Specifically, there are measures of the tax increase and tax cut included in the Special Taxation Measures Law of the national tax and in the preferential measure provided in local tax law.

45 Wada (1992), pp. 74–5, 100–1.

References

Abe, Y. and Awaji, T. (2006) *Environmental Law*, 3rd Edition, Yuhikaku. (In Japanese)

Aoki, K. and Suzuki, N. (2007) 'Analyses on the policy making processes of carbon tax in Japan (1) (2): Political implication of institutional difficulties in the design and introduction', *The Journal of Chiba University of Commerce*, 45.1: 31–43, 42.2: 125–47. (In Japanese)

Fujitani, T. (2010) 'An environmental tax and a provisional tax rate', *Jurist*, 1397: 28–36. (In Japanese)

Hayashi, T. (2001) 'Environmental policy and financial expenditure of local governments', *Review of Agricultural Economics of Hokkaido University*, 57: 1–9. (In Japanese)

Ito, K. (2010) 'Policy process and change: greening of automobile tax in Japan', *Public Policy Studies*, 9: 132–42. (In Japanese)

Jinno, N. (2013) 'Approach from various points of view against an environmental tax', *Local Tax*, 64.4: 2–10. (In Japanese)

Lee, S. (2004) *Theory and Practice of Environmental Subsidies*, The University of Nagoya Press, Japan. (In Japanese)

Ministry of Economy, Trade and Industry. (2011, 2012) *Handbook of Business Tax System*, Research Institute of Economy, Trade and Industry. (In Japanese)

Ministry of the Environment, *Annual Report on the Environment*, Various Issues. (In Japanese)

Ministry of Finance. (2014) *Report on the Result of the Application Survey of the Special Taxation Measures*, Submitted to the 186th Session of the Diet. (In Japanese)

Ministry of Internal Affairs and Communications, *Annual Report of Local Finances*, Various Issues. (In Japanese)

Ministry of Internal Affairs and Communications. (2014) *Report on the Application Status of the Local Tax Burden Reduction Measures, etc.*, Submitted to the 186th Session of the Diet. (In Japanese)

Miyajima, H. (1993) 'Study on environmental tax (carbon tax) through a tax theory', in H. Ishi (Ed.), *Environmental Tax: Practice and Mechanism*, TOYO KEIZAI, 31–46. (In Japanese)

Morotomi, T. (2002a) 'Environmental management by local environmental taxes: their theoretical basis and policy design', *Economia*, 53.1: 43–74. (In Japanese)

Morotomi, T. (2002b) 'Environmental protection cost and burden principle', in S. Teranishi and H. Ishi (Eds.), *Environmental Protection and Public Policy*, Iwanami Shoten, 123–50. (In Japanese)

OECD. (1977) *Environmental Policies in Japan*, Washington, D.C.

OECD. (1994) *OECD Environmental Performance Reviews: Japan*, Paris.

OECD. (2001) *Environmentally Related Taxes in OECD Countries: Issues and Strategies*, Paris.

OECD. (2010) *OECD Environmental Performance Reviews: Japan*, Paris.

Ogata, T. (2006) 'Environmental administration in Japan and the role of local governments', *Papers on the Local Governments System and its Implementation in Selected Fields in Japan*, No.7, Council of Local Authorities for International Relations, Tokyo. (In Japanese)

Takai, T. (2013) *A Theory and Practice of Original Taxation by the Local Government*, Nihon Keizai Hyouronsha. (In Japanese)

Ueta, K. (2002) 'Administrative and financial systems and environmental policy', in S. Teranishi and H. Ishi (eds.), *Public Policy and Environmental Protection*, Iwanami Shoten, 93–142. (In Japanese)

Yokoyama, A. (2014) 'Local environmental taxes: theoretical justification and practice', *The Commercial Review of Chuo University*, 55.18: 30–48. (In Japanese)

Yoshimura, M. (2008) 'Source of revenue for specific expenditure: is the relation between a benefit and a burden necessary?' *Jurist*, 1363. (In Japanese)

Wada, Y. (1992) *Special Taxation Measures*, Yuhikaku. (In Japanese)

Part II
Rebate program for energy-efficient appliances

The Japanese electronic appliance market

Minoru Morita

1. The home appliance market in Japan

"Home appliances" is the general term given to mechanical devices that require electric power and that are used primarily in the residential setting. The Association for Electric Home Appliances (AEHA) classifies home appliances into four categories: "electric equipment," "video and audio equipment," "information and communication equipment," and "other equipment."[1]

Figure II.1 shows the number of home appliances produced and shipped in Japan from 1980 through 2012. According to this figure, in 2012 the total production value of home appliances was 4,466 billion yen ($37.22 billion).[2] In addition, the total value of these shipments was 5,560 billion yen ($46.33 billion). Therefore, the market size of home appliances was equivalent to roughly 1 percent of the nominal GDP in Japan.[3]

However, the size of the home appliance market has shown a downward trend since 1990 (see Figure II.1). Although the production value grew to 10,393 billion yen ($86.6 billion) in 1991, it has been on a gradual decline since that time, dropping to below half of the 1991 level in 2012 (AEHA 2013). This can be attributed to a dramatic strengthening of the yen by the "Plaza Accord" in 1985 and to many Japanese home appliance manufacturers relocating their production facilities to developing countries, which provides the possibility of minimizing production costs and capitalizing on new demand (AEHA 2013).

Figure II.2 shows the production value to exports ratios, the shipment value to imports ratios, and JPY–USD exchange rates. This figure demonstrates that the exchange rate has risen rapidly, from 238.5 yen per dollar in 1985 to 144.8 yen per dollar in 1990. Herewith, the production value to exports ratio rose until 1985, where it peaked at 54.3 percent and then dropped precipitously to 23.5 percent in 1995, ending up at 33.6 percent in 2012. The shipment value to imports ratio, in contrast, increased dramatically from 1.3 percent in 1985 to 29.0 percent in 2012. One can thus infer that, alongside a rapid appreciation of the yen since 1985, Japanese home appliance manufacturers steadily expanded the import and overseas production of home appliances while at the same time scaling back production in Japan.[4]

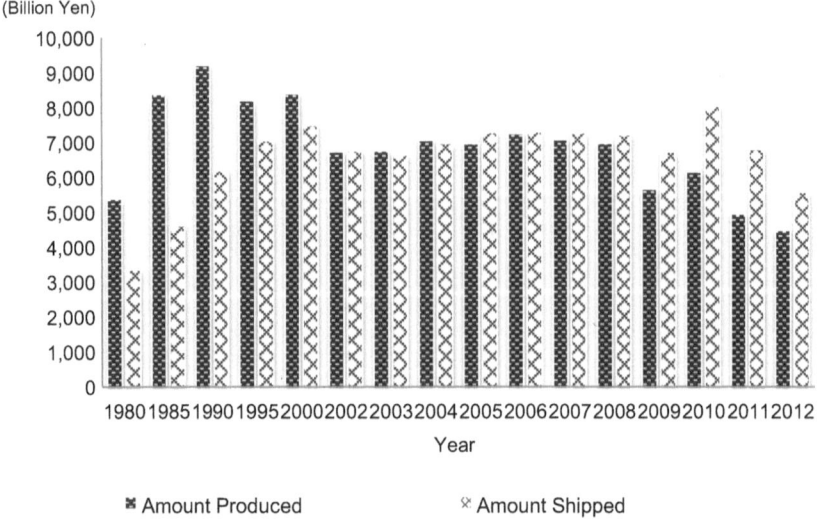

Figure II.1 Production and shipment of home appliances in Japan
Source: Author, based on data from the AEHA (2013).

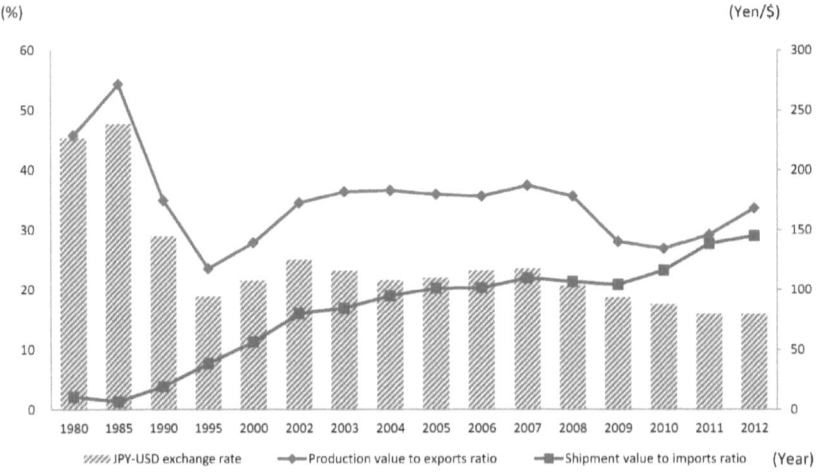

Figure II.2 Production value to exports ratio, shipment value to imports ratio, and exchange rates

Source: "JPY–USD exchange rates" from the Bank of Japan (http://www.stat-search.boj. or.jp/); "Production value to exports ratio" and "Shipment value to imports ratio" calculated from the AEHA (2013).

This trend also applies to the domestic shipment value for home appliances which, as with the domestic production value, is on the decline. The shipment values continued to increase until 2000 (7,474 billion yen), but then went into a gradual decline, hitting 5,560 billion yen in 2012.[5]

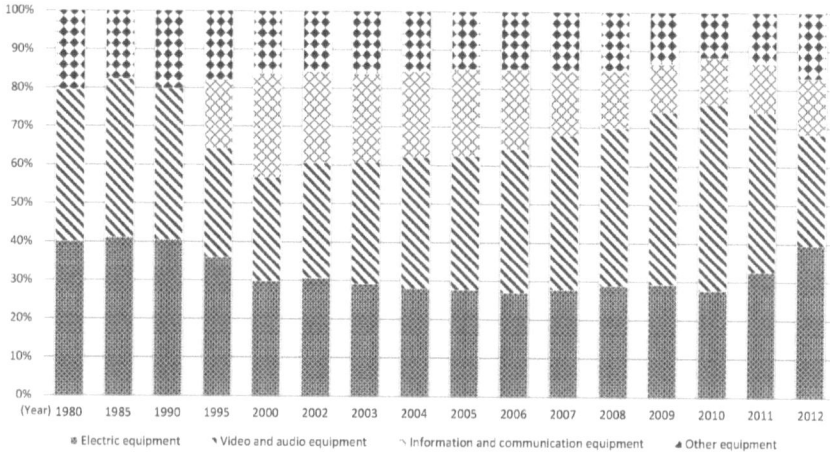

Figure II.3 Share of each home appliance type based on total value of shipments
Source: Author, based on data from the AEHA (2013).

One major factor behind this phenomenon could be that the diffusion of home appliances in Japan was in a period of maturation. Demand for home appliances can be divided into the purchase of a new appliance and the replacement of an old appliance. In the first stage, when home appliance diffusion is low, new product demand increases. However, once the diffusion rate exceeds a certain level, new product purchases drop off and demand for replacement products increases. The rate of increase in demand for home appliances consequently decreases, and the overall market contracts.

Figure II.3 shows the share of each home appliance type that accounts for the total value of electronics shipments in Japan. As shown in this figure, throughout all time periods, electric equipment and video and audio equipment account for more than 60 percent of the total shipment.

Figures II.4.1 and II.4.2 compare the percentage of equipment shipped for the electric equipment and video and audio equipment categories between 1980 and 2012. It is clear from Figure II.4.1 that the percentage of domestic shipment value for electric equipment is at a similar level in 1980 and 2012. Air conditioners account for the largest share (about 30 percent), followed by refrigerators (about 20 percent) and washing machines (10 percent).

In contrast, Figure II.4.2 shows video and audio equipment as a percentage of domestic shipment value. As shown, the 2012 value was much different from that in 1980. In 1980, televisions comprised 40 percent of video and audio equipment, with other products accounting for the remaining 60 percent. In 2012, however, the domestic shipment value included high percentages of new products in the form of digital equipment such as car navigation systems (28 percent) and digital still cameras (10 percent). Televisions accounted for 30 percent of the category in 2012, with consumers moving from CRT televisions to LCD and other digital televisions.

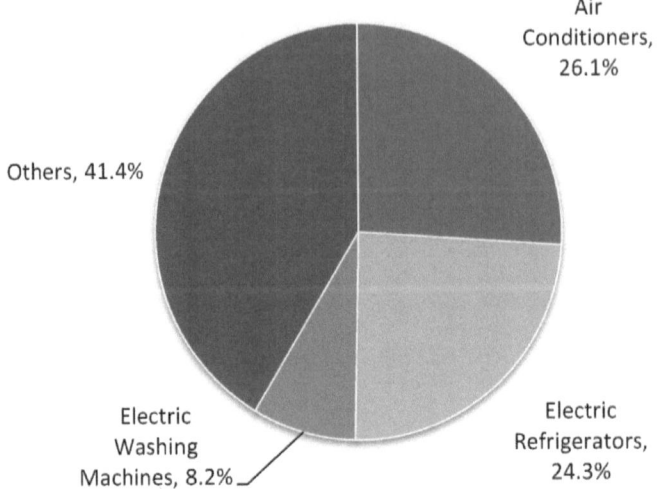

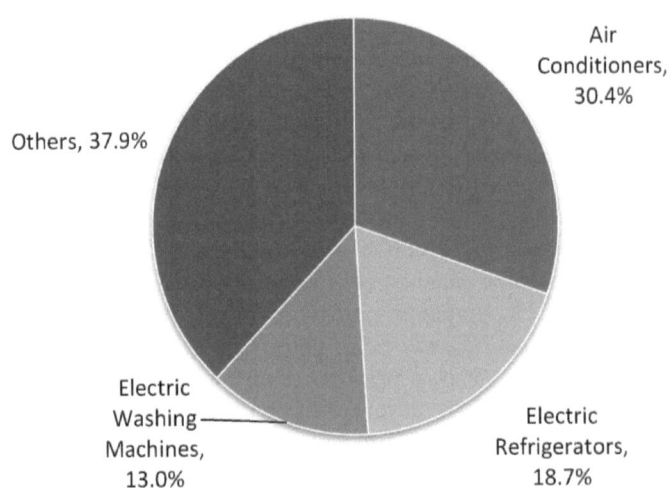

Figure II.4.1 Share of electric equipment based on total value of shipments (Top graph: 1980, bottom graph: 2012)

Source: Author, based on data from the AEHA (2013).

Finally, let us consider whether Japan's home appliance market is currently in a period of maturation. Looking at the diffusion rates in 2000 for major equipment, air conditioners were 86.2 percent, refrigerators 98.0 percent, washing machines 99.3 percent, and TVs 99.0 percent (Energy Data and Modeling Center [EDMC] 2013). The diffusion rates of these major home appliance were over 86 percent, indicating that many households already owned

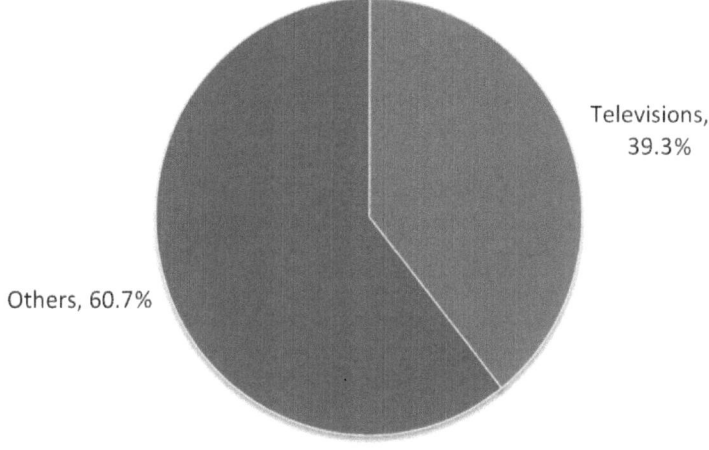

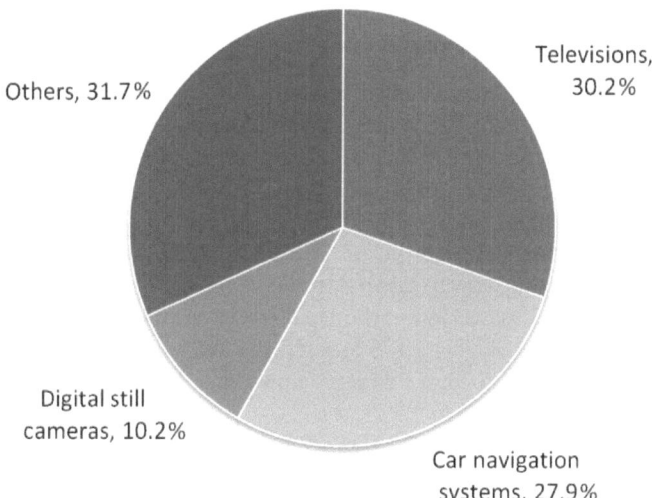

Figure II.4.2 Share of video and audio equipment based on total value of shipments (Top graph: 1980, bottom graph: 2012)

Source: Author, based on data from the AEHA (2013).

such equipment. It is thus possible that Japan's home appliances market is in a period of maturation, with replacement product purchasing being greater than new product purchasing in terms of domestic demand. And, as shown in Figure II.1, the total value of shipments of home appliances in Japan is trending downward.

As noted above, Japan's home appliance market is in decline. Therefore, Japanese home appliance manufacturers are breaking into new markets in developing countries where demand is expected to grow, while at the same time focusing on developing high added-value products in Japan through technological innovation (AEHA 2013).

One of these innovations is "energy efficiency." Improving the energy efficiency of home appliances reduces energy usage. Consumers who replace their existing equipment with energy-efficient products can reduce their utility costs. In addition, the electrical power supply problems that resulted in the aftermath of the Great East Japan Earthquake that occurred in March 2011 further heightened Japan's desire to conserve energy. In light of this, Japanese home appliance manufacturers are working to develop energy-saving products and bring them into wider use. In connection with its efforts to combat global warming, the Japanese government is also taking a number of measures aimed at promoting better home appliance energy efficiency and wider use of energy-efficient products.

The following section looks at the importance of energy efficiency for home appliances and the measures the Japanese government has taken to promote the use of energy-efficient appliances.

2. Energy Conservation Act

The Japanese government has taken a number of measures to promote energy efficiency based on the Law Concerning the Rational Use of Energy, or the "Energy Conservation Act," which was enacted in 1979. The purpose of this act was to decrease the use of electric power and fuel in all sectors (industry, residential, commercial, and transportation) in Japan (Nishiyama 2013).

The act was passed during the oil crisis of the 1970s.[6] This crisis dealt a considerable economic blow to Japan due to the country's significant dependence on foreign energy sources. Hence, the Japanese government saw the importance of promoting energy efficiency and enacted the Energy Conservation Act with the goal of balancing domestic energy supply and demand and guaranteeing energy security.

Table II.1 presents a summary of the Energy Conservation Act by economic sector. When the act was first established, the primary targets for regulation were the industrial and commercial sectors. Factories and offices that used more than a predefined amount of energy (specifically, energy consumption more than 1,500 kl) were designated as "energy management factories."[7] Energy managers were assigned to these facilities, which were required to produce regular energy usage reports and submit medium- to long-term plans concerning energy conservation. The result was systematic and voluntary enforcement of energy management, particularly in the manufacturing industry.

Since 1979, a number of amendments have been added to the Energy Conservation Act. The primary reasons for the amendments were to address increases in energy consumption and to strengthen global warming mitigation measures to combat climate change. For example, owners of buildings with more than 300 m^2 of floor space and home sellers who sold 150 houses or more a year were

Table II.1 A summary of the Energy Conservation Act

Sector	Target	Note
Industrial/ Commercial Sector	Designated factories/ offices (energy usage of 1,500 kl/year or more)	Assign an energy manager Submit reports on energy usage
Industrial/ Commercial/ Residential Sector	Designated buildings (floor area of 300 m² or more)	Notify of energy conservation measures
	Designated housing suppliers (150 or more properties a year)	Meet home energy efficiency performance targets
Residential/ Commercial Sector	Designated goods (cars, equipment)	Improve energy efficiency level (Top Runner program)
	Equipment retailers	Display energy efficiency
	Energy suppliers	Provide the information of energy efficiency in store
		Promote energy efficiency apparatuses
		Provide energy-saving devices
Transportation Sector	Designated transport businesses (those with 200 or more trucks and 300 or more trains)	Submit reports on energy usage
	Designated shippers (those with shipping amounts of 30 million kt or more a year)	

Source: Nishiyama (2013) and EDMC (2014).

targeted for energy efficiency regulations. Energy efficiency regulations were also implemented in the transportation sector. Passenger and freight carriers that owned more than 200 trucks or more than 300 railroad carriages were obliged to report their energy use.

Of course, the residential sector was not exempt from regulation. In the residential sector, measures were enacted to improve the energy efficiency of the home appliances. Specifically, the "Top Runner program" regulates the energy efficiency of home appliances and automobiles. In addition to regulations, a subsidy and rebate program – the "eco-point program" (EPP) – was implemented. Thus, for the residential sector, home appliances have been the primary target of energy efficiency regulations. Because of this, the next section explains energy efficiency policies affecting the residential sector.

3. Energy efficiency measures for home appliances

Energy efficiency measures targeting the residential sector have centered on regulations aimed at improving the energy efficiency of products such as home appliances and automobiles (EDMC 2014). A core element of these measures is the Top Runner program. In addition, an environmental labeling system has also been established that mandates the display of energy efficiency information for home appliances for consumers, with the aim of promoting the diffusion of energy-efficient products.

For household goods such as home appliances and automobiles, the Top Runner program selects the most energy-efficient products currently on the market to serve as standards and then mandates that all manufacturers manufacture products that meet these standards by a specific target year. Specifically, import businesses and manufacturers of energy-consuming products such as home appliances and automobiles are required to meet Top Runner standards by a target year, usually 3 to 10 years out (Ministry of Economy, Trade and Industry and Agency for Natural Resources and Energy 2010). Businesses that are unable to satisfy Top Runner standards by the target year are required to state their reasons for not meeting the standards and to report on the measures they will take to achieve compliance. The Ministry of Economy, Trade and Industry makes recommendations to noncompliant businesses to improve their performance, goes public with the recommendation, and issues fines of up to 1 million yen ($8,333.33).

Table II.2 shows the product categories targeted by the Top Runner program. As of 2013, there were 26 product categories, comprised mostly of home appliances

Table II.2 Product categories targeted by the Top Runner program

Target Products	Target Products	Target Products
1) Automobiles	10) Electric refrigerators	19) Jar rice cookers
2) Trucks	11) Electric freezers	20) Microwaves
3) Air conditioners	12) Stoves	21) DVD recorders
4) TV receivers	13) Gas cooking appliances	22) Routing devices
5) Videotape recorders		23) Switching devices
6) Lighting fixtures	14) Gas water heaters	24) Multifunction devices
7) Copy machines	15) Kerosene water heaters	25) Printers
8) Computers	16) Bidets	26) Electric water heaters (heat-pump water heaters)
9) Magnetic disk devices	17) Vending machines	
	18) Transformers	

Source: Ministry of Economy, Trade and Industry and Agency for Natural Resources and Energy (2010).

Table II.3 Energy efficiency improvements as a result of the Top Runner program

Target Devices	Average Energy Efficiency Improvement (Actual)
Air conditioners (household models rated 4 kW or lower)	67.8% (1997 to 2004 comparison) 16.3% (2005 to 2010 comparison)
TVs	29.6% (2004 to 2008 comparison)
Electric refrigerators	55.2% (1998 to 2004 comparison) 43.0% (2005 to 2010 comparison)
Fluorescent lighting	35.7% (1997 to 2005 comparison)
Computers	99.1% (1997 to 2005 comparison) 80.8% (2001 to 2007 comparison)

Source: Ministry of Economy, Trade and Industry and Agency for Natural Resources and Energy (2010).

(Agency of Natural Resources and Energy 2013). Table II.3 shows the energy efficiency improvements achieved for products through the Top Runner program. According to this table, air conditioners, for example, saw an energy efficiency improvement of 67.8 percent compared to the average energy efficiency improvement of products of the same size (household air conditioning units of 4 kW or less) when comparing 2004 to 1997, the year the Top Runner program was established (Nishiyama 2013). Refrigerator efficiency increased 55.2 percent from 1998, the year the Top Runner program was established for this category, to 2004. As we can see, the Top Runner program is gradually affecting energy efficiency improvements for home appliances.

In Japan, the Top Runner program is being carried out in conjunction with other environmental-labeling systems, such as the "Energy-Saving Labeling System" and the "Uniform Energy-Saving Label Program" (Agency of Natural Resources and Energy 2011). Labeling programs mandate the labeling of various home appliances to enable consumers to easily understand their energy efficiency. The Energy-Saving Labeling System was begun in 2000 to promote the diffusion of energy-efficient products by requiring that home appliance manufacturers provide consumers with information (including energy efficiency standard achievement rates, annual power usage, etc.) concerning the energy efficiency of their products (see Figure II.5). Established through an April 2006 amendment to the Energy Conservation Act, the Uniform Energy Efficiency Label Program requires retailers to provide data on the energy efficiency of various home appliances. This program uses a five-star rating system to make it easy for consumers to understand how energy efficient a particular product is (see Figure II.5). Because of the assistance they provide in helping

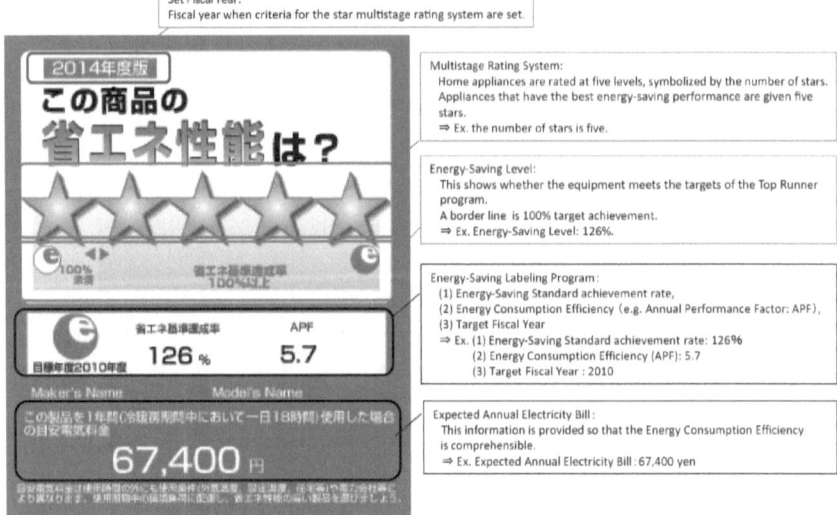

Figure II.5 Uniform Energy-Saving Label Program (air conditioner)

Source: Ministry of Economy, Trade and Industry and Agency for Natural Resources and Energy (2010) and Agency for Natural Resources and Energy (2014).

consumers understand the energy efficiency of different products, these programs are thought to be effective in improving the diffusion of energy-efficient products.

In addition to the energy efficiency measures mandated by the aforementioned amendments to the Energy Conservation Act, Japan also conducted a rebate program called the "eco-point program" (EPP). The EPP was conducted from 15 May 2009 to the end of March 2011 and targeted air conditioners, refrigerators, and televisions. Under the EPP, consumers who purchased energy-efficient products were given points that they could redeem for various goods and services. Table II.4 shows the points granted by type product. From this, we can see that products with higher capacities garnered more points. In the case of air conditioners, for example, consumers received 6,000 points for products rated 2.2 kW or lower and 9,000 points for larger products rated 3.6 kW or higher.

The EPP was implemented for three main reasons. The first was as a measure against global warming. The rate of CO_2 emissions had increased throughout the entire consumer sector, including the household sector, and had much higher increases than other economic sectors. The greatest contributor to CO_2 emissions in the consumer sector was residential energy use. It was thus necessary to reduce energy usage and decrease CO_2 emissions in this sector by promoting purchase of high-energy-efficiency products to replace low-energy-efficiency products.

Table II.4 Number of points granted according to size

	Televisions Points			Refrigerators Points			Air Conditioners Points		
Capacity	Before Revision	After Revision	Capacity	Before Revision	After Revision	Capacity	Before Revision	After Revision
46 V or more	36,000	17,000	501 L or more	10,000	5,000	3.6 kW or more	9,000	5,000
42 V to 40 V	23,000	11,000	401 to 500 L	9,000	5,000	2.8 kW to 2.5 kW	7,000	4,000
37v	17,000	8,000	251 to 400 L	6,000	3,000	2.2 kW or less	6,000	3,000
32 V to 26 V	12,000	6,000	250 L or less	3,000	2,000			
Less than 26 V	7,000	4,000						

Note: The "Points after Revision" refers to the number of points granted to consumers who purchased products after the program revision on 1 December 2010 and before 31 March 2011.

Source: Author, based on data from the Ministry of the Environment (http://www.env.go.jp/policy/ep_kaden/buy/point.html).

The second reason was as an economic measure to mitigate the recession caused by the financial crisis that began in late 2008. The EPP brought an economic injection of roughly 693 billion yen ($5.78 billion) throughout the period of the program. According to government calculations, the number of target units shipped increased 67 percent and sales increased by some 2.6 trillion yen ($21.67 billion). This had an economic effect of around 5 trillion yen ($41.67 billion) and resulted in the creation or retention of approximately 320,000 jobs (Ministry of the Environment, Ministry of Economy, Trade and Industry, and Ministry of Internal Affairs and Communications, 2011).

The third reason was to complete the transition to terrestrial digital broadcasting. At the time there was a need to promote the diffusion of televisions capable of digital broadcasting in association with the start of terrestrial digital broadcasting. In fact, of the three product categories, televisions had the highest number of replacement purchases.

This part discusses the impact of measures that have been used to promote the manufacture and purchase of energy-efficient appliances in Japan.

Acknowledgment

This chapter is sponsored by the financial aid from the Center for Global Partnership, Japan Foundation.

Notes

1 The categories listed here were determined by the AEHA (2013): (1) "electric equipment" includes products such as electric refrigerators, air conditioners, electric washing machines, freezers, dishwashers, dryers, microwaves, IH cooking heaters, IH cooking heaters, jar rice heaters, electric clothes dryers, electric vacuum cleaners, electric shavers, hairdryers, ventilation fans, electric fans, air purifiers, dehumidifiers, and bidets. (2) "Video and audio equipment" includes televisions (cathode-ray tube TVs, LCD TVs, plasma TVs), digital still cameras, radios, video production equipment, VTR/DVD/BR recorders and players, camcorders, stereo systems, components, digital audio players and other sound equipment, as well as car navigation systems, and other audiovisual equipment for cars. (3) "Information and communication equipment" includes personal computers and mobile phones (cell phones and PHS devices), telephones, and fax machines. (4)"Other equipment" includes lighting equipment such as home lighting fixtures and LED lamps, as well as light bulbs and batteries such as dry cell batteries and storage batteries.
2 120 yen is converted into $1.
3 Japan's nominal GDP in 2012 was 473,777 billion yen ($3,948.14 billion).
4 Figure II.2 also shows that although the value to exports ratio has been on the rise since 2010, this upward trend is deceptive, as there was actually a dramatic decrease in the domestic output value for audiovisual equipment as a part of home appliances (AEHA, 2013).
5 One thing to note is that the domestic shipment value went up in 2010. This is attributed to the "eco-point program," which was executed by the Japanese government from May 2009 to March 2011 as a measure against global warming and which resulted in a temporary rise in replacement product purchasing. The eco-point program is discussed in Chapter 5.

6 The first and second oil crises were in 1973 and 1979, respectively.
7 Parts of the commercial sector, such as office buildings and hotels, were also included in the regulations.

References

Association for Electric Home Appliances. (2013) 'Handbook of Kaden-Sangyou', Association for Electric Home Appliances publication (In Japanese).

Agency of Natural Resources and Energy. (2011) 'Wagakuni no Shoenerugi-seisaku ni tu i te', Available: <http://www.enecho.meti.go.jp/category/saving_and_new/saving/pdf/current_situation_japanese.pdf/> (accessed 9 August 2014) (In Japanese).

Agency for Natural Resources and Energy. (2013) 'Kongo-no Shoenerugi-seisaku ni tu i te', Available: < http://www.chubu.meti.go.jp/enetai/shouene/kouhou-setumeikai/25fy/130919gaiyou/1.pdf/> (accessed 9 August 2014) (In Japanese).

Agency for Natural Resources and Energy. (2014) 'Shoene-gata Saihin Jyohou', Available: <http://seihinjyoho.go.jp/> (accessed 9 August 2014) (In Japanese).

Energy Data and Modeling Center. (2013) 'Handbook of Energy and Economic Statistics in Japan', The Energy Conservation Center Japan publication.

Energy Data and Modeling Center. (2014) 'Japan Energy Conservation Handbook 2013', Energy Conservation Center Japan publication.

Ministry of Economy, Trade and Industry and Agency for Natural Resources and Energy. (2010) 'Top Runner Program', Available: <http://www.enecho.meti.go.jp/category/saving_and_new/saving/003/> (accessed 9 August 2014).

Ministry of the Environment, Ministry of Economy, Trade and Industry and Ministry of Internal Affairs and Communications. (2011) *Effects of the Home Appliance Eco-Point System Policy*, Available: <http://www.meti.go.jp/press/2011/06/20110614002/20110614002–2.pdf> (accessed 1 October 2014) (In Japanese).

H. Nishiyama, (2013) 'Japan's Policy on Energy Conservation', EMAK 4th Work Shop, Available: <http://eneken.ieej.or.jp/data/4749.pdf> (accessed 9 August 2014).

4 Effect of an eco-point program on consumer digital TV selection

Keiko Yamaguchi, Shigeru Matsumoto, and Tomohiro Tasaki

4.1. Introduction

The switch from analog to digital broadcasting has been a global movement. The United States (U.S.) completed the transition from analog to digital television broadcasting on Friday, 12 June 2009. The European Union (EU) recommended the end of 2012 as the deadline for switching to analog terrestrial television, and 22 of the 27 EU member states implemented their transition in line with this recommendation (European Audiovisual Observatory 2012).

Japan terminated analog broadcasting and completed the switch to digital terrestrial television broadcasting on Sunday, 24 July 2011 (Ministry of Internal Affairs and Communications [MIAC] 2011). To accelerate the transition to digital broadcasting, the Japanese government implemented an incentive-based program called the "eco-point program for appliances" (EPP). During EPP periods, consumers who purchased a digital television (TV) obtained eco points (rebates) that could be used to buy other goods and services. By providing eco points only for energy-efficient digital TVs, the Japanese government also sought to persuade manufacturers to improve the energy efficiency of digital TVs.

A variety of economic instruments have been used to reduce household energy consumption, including energy taxes, grants, preferential loans to invest in energy-efficient equipment, and financial incentives to promote the installation of solar panels and residential wind turbines (OECD 2011).

Past studies have found that consumers tend to underappreciate the benefits of future energy savings, and therefore underinvest in energy-saving technologies. To alter this myopic behavior, more and more governments are implementing rebate programs. In recent years, consumers have been able to receive rebates for a variety of products, such as energy-efficient appliances, houses, hybrid vehicles, and solar panels. Previous studies have reported that rebate programs provide consumers with a stronger incentive than traditional tax-deduction approaches and rapidly expand the market penetration of energy-efficient products (e.g. Gallagher and Muehlegger 2011).

When rebate programs are used to promote energy-efficient products, consumers obtain rebates only if they purchase products that qualify as being energy efficient. Rebate programs decrease the acquisition cost for energy-efficient products compared to energy-inefficient ones, thereby inducing consumers to choose energy-efficient products. In addition to this substitution effect, rebate programs give people more money to spend, thus allowing them to purchase more expensive, larger products.[1] Therefore, the income expansion effect must be taken into account in order to assess the overall impact of a rebate program.

Khazzoom (1980) argued that an increase in the energy efficiency of appliances would increase energy demand. An improvement in energy efficiency lifts real income, which, in turn, causes consumers to increase their energy consumption. The higher energy demand caused by energy efficiency improvements is called the "rebound effect," and has been studied extensively over the last several decades in the context of residential energy demand and automobile travel.

Empirical studies such as those by Dubin, Miedema, and Chandran (1986), Greening, Greene, and Difiglio (2000), Hausman (1979), and Schwarz and Taylor (1995) confirmed the rebound effect in the context of residential energy demand and found that an increase in energy efficiency was partly offset by higher household energy consumption.

Technology improvements have reduced gasoline consumption per unit distance. However, as fuel economy improves, travel demand also increases. The rebound effect of automobile travel has been confirmed in previous papers. According to Haughton and Sarkar (1996) and Jones (1993), improvements in fuel economy have increased the number of vehicle miles traveled by 10 to 30 percent.

Although many scholars have studied the income expansion effect on energy consumption in the context of the rebound effect caused by technology improvements, none have analyzed the income expansion effect of rebate programs.[2]

Rebate programs allow consumers to purchase expensive products that would otherwise be unaffordable without a rebate. Since many rebate programs use product price as the basis of the rebate, consumers receive larger rebates if they purchase a more expensive product. Many of these expensive products are large and have many complex functions, and therefore consume more energy than less expensive products.

In this chapter, we evaluate the impact of the EPP on the energy consumption of digital TVs by determining whether the EPP induced consumers to purchase energy-efficient TVs or large-screen TVs and how much electricity was saved by the program.

The rest of the chapter is organized as follows. In Section 4.2, we explain Japan's digital broadcast transition and eco-point program for appliances. In Section 4.3, we discuss the data and specify the empirical model; we employ a conditional logit model to characterize consumers' selection of digital TVs. Section 4.4 presents our empirical findings, which demonstrate that the EPP expanded the market share of energy-efficient TVs within a specific screen-size category but also induced consumers to purchase larger TVs through the income expansion effect. Using data on the annual energy consumption of TVs, we show that the EPP increased energy consumption from TV usage. Section 4.5 concludes the chapter.

4.2. Digital broadcast transition and the eco-point program

The transition from analog to digital television is regarded as the most significant advance in television technology since the introduction of color TV. Digital broadcasting offers better image quality and sound than analog TV, while also allowing households to access more channels and to connect televisions to household networks.

This transition to digital television is a worldwide trend. Most developed countries have already completed the transition, and some developing countries are now moving in the same direction (Tokyo International Communication Committee 2010).

The Japanese government had set a goal of completing Japan's nationwide transition to digital broadcasting by 24 July 2011, but found that progress was slow. A survey conducted by the MIAC (2009) revealed that by March 2009 only 69.5 percent of receivers were digital. To accelerate the transition, the government introduced an incentive-based program called the eco-point program (EPP).

In addition to accelerating the digital broadcasting transition, the government had two other objectives for the EPP: (1) to reduce CO_2 emissions by encouraging consumers to choose energy-saving electric appliances and (2) to mitigate the negative impact of the Lehman shock by stimulating domestic consumption. Air conditioners and refrigerators were also covered by the EPP.

Figure 4.1 summarizes the timeline of key events in the EPP. The program started on 15 May 2009 and ended on 31 March 2011. During this period, the program was modified several times: for example, the number of eco points provided was reduced on 30 November 2010, and the energy efficiency criteria were tightened on 31 December 2010. In addition to eco points, recycle points were provided to consumers who replaced old appliances by 31 December 2010.

The impact of the program on the domestic economy was significant. The Ministry of Economy, Trade and Industry (2011) estimated that the shipment of household electric appliances increased by 67 percent and that the program expanded the domestic economy by 2.6 trillion yen ($21.67 billion) while it was in effect.[3]

Figure 4.2 shows the percent change from the previous year of domestic digital TV shipments. Because digital TVs are a new and emerging technology, the

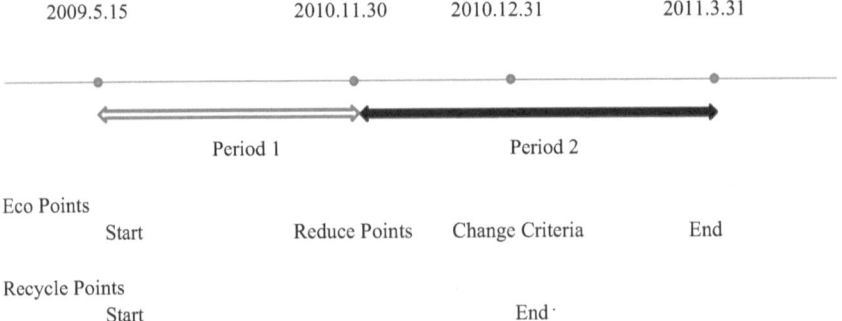

Figure 4.1 Timeline of key events in the eco-point program

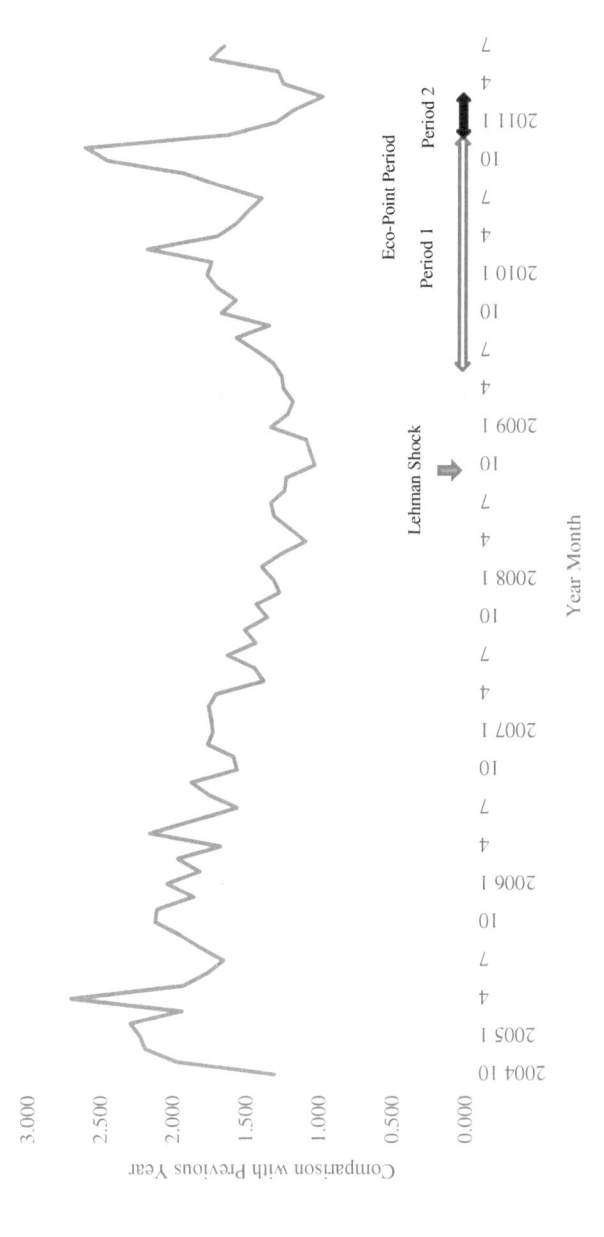

Figure 4.2 Change in digital TV shipments

(Source: Japan Electronics and Information Technology Industries Association 2011)

number of shipments expanded over the sample period. However, Figure 4.2 shows that the speed of the expansion decelerated beginning in April 2005 and that the introduction of the EPP in May 2009 reversed this trend, reaccelerating the growth of digital TV shipments.

The rebate points that a consumer could obtain through the EPP varied by appliance type and size (see Table II.4 on page 73). More eco points were provided for digital TVs with larger screens. For instance, if a consumer purchased a digital television with a screen size of less than 26 inches, s/he only received 7,000 yen ($58.33), in contrast to the receipt of 36,000 yen ($300) for the purchase of a digital television with a screen size greater than 46 inches.

Because a large-screen TV had a larger rebate than a small-screen TV, we expected the incentive for consumers to choose an energy-efficient TV to be stronger for large-screen TVs than for small-screen TVs. However, we also expected EPP-induced consumers to choose a large-screen model because of the greater benefit of a large-screen TV relative to a small-screen TV. Therefore, to assess the overall impact of the EPP, we needed to estimate these two effects of the EPP.

4.3. Data and empirical model

4.3.1. Data

The primary data used in our analysis were sales data collected by the market research firm GfK Marketing Service Ltd. Japan, which is a member of the GfK group, one of the largest market research firms in the world, headquartered in Nuremberg, Germany.

GfK Marketing Service Ltd. Japan conducted a survey about household electric appliances, cameras, and telecommunication equipment. Data on TVs were collected from 4,023 retail stores and covered about 75 percent of total annual sales in Japan.

The data collected provided the total value and number of sales for each product month. We calculated the average sales price of each product by dividing the total value of sales by the total number of sales, and this average sales price was used in the following analysis. In addition to price information, the data contained the detailed specifications of each product, which we also included in our analysis.

We classified digital TVs into three energy-efficient classes according to the eco-point classification: non-eco TVs, for which consumers could not obtain any eco points; eco TVs, which attracted eco points only in period 1; and super-eco TVs, which attracted eco points in both periods 1 and 2. Figure 4.3 describes the change in market share across the three eco classes, and shows that the market share of non-eco TVs decreased during the whole sample period while that of super-eco TVs increased. For eco TVs, we observed an inverse U-shape relationship, whereby market share increased in the initial period and then decreased in the latter period. Overall, the figure shows that sales of energy-efficient digital TVs increased.

Figure 4.3 Change in market share of digital TVs

In this chapter, we focus on TVs whose screen sizes ranged from 16 to 61 inches, which included 98 percent of total TV sales, and classify TVs into eight screen size classes. Table 4.1 compares the price and market share across different screen-size classes. Screen size is the most important determinant of the price of digital TVs, and the table clearly shows that the price of a digital TV increases as the screen size increases. For instance, the mean price of eco TVs is 47,992.1 yen in screen size class 1, while it is 292,337.7 yen in screen size class 8.

We expected the price of non-eco TVs to be lower than that of eco or super-eco TVs within the same screen-size class. However, Table 4.2 shows that this

Table 4.1 Summary statistics of digital television sales

Screen Size Class	Eco Class	Price (yen)[a]		Sale	
		Mean	Standard Deviation	Number of Sales	Market Share
1	1. Non-eco	33,846.6	10,533.7	677,595	1.88%
(16–20)	2. Eco	47,992.1	15,571.4	3,670,607	10.16%
	3. Super-eco	33,561.8	5,290.5	1,902,941	5.27%
2	1. Non-eco	37,399.4	8,187.8	158,074	0.44%
(21–25)	2. Eco	58,745.5	31,942.8	643,363	1.78%
	3. Super-eco	47,877.1	6,208.9	1,163,718	3.22%
3	1. Non-eco	56,026.5	19,310.8	64,314	0.18%
(26–31)	2. Eco	74,756.3	17,985.6	1,970,627	5.46%
	3. Super-eco	63,071.8	7,461.2	1,118,407	3.10%
4	1. Non-eco	67,390.1	33,174.7	149,253	0.41%
(32–36)	2. Eco	81,215.1	23,596.4	6,932,429	19.19%
	3. Super-eco	90,375.8	30,518.9	5,228,310	14.48%
5	1. Non-eco	10,5865.9	36,671.5	43,389	0.12%
(37–39)	2. Eco	127,329.2	29,271.7	2,346,698	6.50%
	3. Super-eco	128,957.8	33,981.8	907,003	2.51%
6	1. Non-eco	129,590.2	54,998.6	240,748	0.67%
(40–45)	2. Eco	146,446.1	43,466.1	3,046,350	8.43%
	3. Super-eco	137,422.6	43,133.4	3,560,032	9.86%
7	1. Non-eco	248,322.5	57,577.0	165,695	0.46%
(46–50)	2. Eco	206,532.7	59,805.2	837,903	2.32%
	3. Super-eco	209,464.2	40,758.8	859,671	2.38%
8	1. Non-eco	437,398.7	124,981.8	14,526	0.04%
(51–61)	2. Eco	292,337.7	48,510.3	202,437	0.56%
	3. Super-eco	283,224.9	51,047.9	215,419	0.60%

a Only data with positive sales numbers are used. $1 is approximately 120 yen.

Table 4.2. Estimation results of the multinomial logit model

Variable	Model 1		Model 2	
	Coefficient	*Std.*	*Coefficient*	*Std.*
Price	−1.237E−05	1.084E−11	−4.389E−06	1.186E−11
Screen	0.03783	5.276E−08	0.13567	3.257E−07
Days after initial sales	−0.00106	6.397E−10	−0.00093	9.547E−10
Control value	−3.06E−05	1.772E−11	9.4744E−07	1.969E−11
Manufacturer dummies				
Manufacturer A	−1.95921	3.672E−06	−4.76968	4.105E−06
Manufacturer B	2.34618	1.978E−06	1.33395	1.975E−06
Manufacturer C	3.08635	1.683E−06	3.19230	1.848E−06
Manufacturer D	1.67177	2.301E−06	2.53068	2.542E−06
Manufacturer E	1.48418	1.887E−06	1.49894	1.805E−06
Function dummies				
Blu-ray dummy	−2.61021	3.609E−05	9.95962	4.403E−05
DVD dummy	5.68545	3.603E−05	−5.81724	4.370E−05
Full HD dummy	−0.37399	7.697E−07	0.45559	1.953E−06
HDD dummy	−0.87933	1.105E−06	−0.82616	1.212E−06
HDMI dummy	0.33077	7.106E−07	0.1151	7.847E−07
Digital churner	−5.61774	4.945E−06	−4.8195	7.392E−06
Screen separation dummy	−1.15964	1.605E−06	−1.34403	1.623E−06
3D dummy	−0.49086	3.504E−06	0.87853	3.399E−06
Screen-size fixed effects				
Size 1			2.17743	5.475E−06
Size 2			0.54766	2.295E−06
Size 3			0.04192	1.183E−06
Size 4			0.24152	5.321E−07
Size 6			−0.08263	2.559E−07
Size 7			−0.38371	4.706E−07
Size 8			−0.65073	7.086E−07

Note: All coefficients are statistically significant at the 1% level.

is not always the case: the mean price of non-eco TVs is higher than for eco or super-eco TVs in several screen classes. This phenomenon occurs because the price of liquid crystal panel (used in modern TVs) decreased during the sampling periods. The market share of non-eco TVs was high when the price of liquid crystal panel was high. In contrast, the market shares of eco and super-eco TVs increased after the price of liquid crystal panel dropped. This explains why the mean price of non-eco TVs was higher than that of eco or super-eco TVs.

4.3.2. Empirical model

The number of types of digital TVs sold during our sampling periods exceeded 45,000, so we needed to reduce the choice set for the empirical analysis. In this study, we classified all TVs into 24 categories (8 screen classes × 3 energy-efficient classes) and then assumed that a consumer would choose from one of these 24 categories when purchasing a TV.

In each category, we averaged out the TV characteristics. Specifically, we calculated the weighted averages of TV characteristics in each category and then created an average TV for each category. We assumed that a consumer took into account the average characteristics of TVs and then chose a TV from one of the 24 categories that would maximize his or her utility.

We employed a conditional logit model to describe the consumer's selection of a TV. The consumer obtains utility u_j if s/he purchases a digital TV in category j at time t. The utility function of the consumer is characterized by the following function:

$$u_j = \gamma_0 + \gamma_p p_{tj}^A + \sum_k \gamma_k x_{tjk} + \varepsilon_{tj} \tag{4.1}$$

where p_{tj}^A denotes the real acquisition price of a digital TV and x_{tjk} denotes the observed product characteristics. We estimated the real acquisition price by subtracting eco points, r_{tj}, from the average sales price, p_{tj}.

The last term, ε_{tj}, is the error term, and it contains an unobserved factor, which is likely to be correlated with the price, p_{tj}. Because of this price endogeneity problem, the direct application of the conditional logit model can produce inconsistent estimates. We resolved the price endogeneity problem with the control function approach (Kim and Petrin 2010; Petrin and Train 2010).

Our control function approach consisted of two stages.[4] In the first stage, we conducted a hedonic price analysis and estimated the expected price. We then calculated the residual for each observation:

$$\mu_{tj} = p_{tj} - E\left[p_j \mid Z_{mj}\right] \tag{4.2}$$

where p is the mean sales price and Z is the vector of product characteristics that include x_{tjk}. The result of the first-stage analysis is presented in Appendix 4.A.

In the second stage, we plugged the residual derived by Equation 4.2 into Equation 4.1 and obtained the following utility function:

$$u_j = \gamma_0 + \gamma_p p_{tj}^A + \sum_k \gamma_k x_{tjk} + \lambda \mu_{tj} + \epsilon_{tj} \tag{4.3}$$

where ϵ is i.i.d logit. We then estimated a conditional logit model using the maximum likelihood method, treating μ_{tj} as an additional repressor.

4.4. Empirical results

4.4.1. Consumer TV selection

Table 4.2 presents the estimation results of the conditional logit models. We estimated two models. Model 1 is the basic model, whereas Model 2 includes screen-size fixed effects. In the table, we report the estimation results in which all variables become statistically significant.

In both models, the price variable becomes negative, implying that an expensive TV is less likely to be chosen if TV functions remain constant. In contrast, the screen variable becomes positive. Therefore, a consumer is more likely to choose a large-screen TV if other variables remain constant. As we would naturally expect, a consumer prefers to purchase a large-screen TV at a low price.

The variable of days after initial sales turns out to be negative in both models, implying that a consumer prefers a newer model. Although we obtained an opposite sign between the two models, the control function variable becomes statistically significant, which indicates that an unobserved factor must be controlled for before applying a conditional logit model.

The estimated coefficients vary across the two models because the number of features on a TV is associated with screen size. Some features are standardized on large-screen TVs but optional on a small-screen TVs.

4.4.2. Impact of the eco-point program

We assumed that the total number of TV sales would remain the same when the EPP was not available. We then estimated the impact of the EPP on sales of digital TVs. The result, presented in Table 4.3, reveals that the EPP reduced sales of non-eco TVs in all screen-size classes. In total, the EPP reduced sales of non-eco TVs by 50,492, and thereby succeeded in driving out energy-inefficient models.

We then divided the change in the number of TV sales by the expected number of TV sales when the EPP was not available. The numbers in parentheses in Table 4.3 indicate the percent level changes, and these show that the impact of the EPP for large-screen classes was greater than that for small-screen classes. This is an expected result, because the amount of the rebate was greater for large-screen classes than for small-screen classes.

Table 4.3 also shows that the EPP increased the sales share of large-screen TVs but decreased the sales share of small-screen TVs. For example, the EPP increased sales of TVs in screen-size class 6 by 163,469, while it decreased sales in screen-size class 4 by 93,668. This difference suggests that the EPP induced consumers to choose large-screen TVs.

Table 4.3 Estimated impacts of eco-point program on digital TV sales

Eco Class	Screen-Size Class								Sum
	1 (16-20)	2 (26-31)	3 (32-36)	4 (32-36)	5 (37-39)	6 (40-45)	7 (46-50)	8 (51-61)	
Non-eco	-6,099 (-2.25%)	-1,777 (-1.61%)	-4,355 (-22.57%)	-6,485 (-15.62%)	-8,463 (-76.17%)	-15,726 (-46.74%)	-5,467 (-20.04%)	-2,121 (-33.06%)	-50,492 (-9.71%)
Eco	-76,247 (-3.01%)	-18,089 (-2.74%)	-22,199 (-1.60%)	-53,615 (-1.12%)	10,709 (0.88%)	62,013 (3.09%)	66,122 (12.14%)	13,172 (12.93%)	-18,134 (-0.14%)
Super-eco	-48,843 (-2.75%)	-30,422 (-2.75%)	-6,835 (-0.66%)	-33,569 (-0.69%)	10,612 (1.26%)	117,182 (3.55%)	47,845 (6.13%)	12,656 (6.59%)	68,626 (0.49%)
Sum	-131,188 (-2.86%)	-50,288 (-2.68%)	-33,389 (-1.36%)	-93,668 (-0.97%)	12,858 (0.62%)	163,469 (3.06%)	108,500 (8.02%)	23,707 (7.90%)	

The GfK data included information on the annual energy consumption of each TV model. From that information, we calculated the average energy consumption of digital TVs in the 24 categories. We then multiplied the estimated change in TV sales by the average energy consumption in order to evaluate the impact of the EPP on aggregate energy consumption. In our estimates, the EPP increased annual electricity consumption from TV usage by 23,975,135 kWh. Class 8 was the most popular screen-size class. The average annual energy consumption of non-eco TVs in this class was about 155 kWh; therefore, the impact of the EPP corresponds to an additional 154,678 in sales of these TVs.

4.5. Conclusion

In this chapter, we examined how the eco-point program – a rebate program recently introduced by the Japanese government – affected consumers' selection of digital TVs. After controlling the price endogeneity problem with a control function approach, we estimated a conditional logit model. Our empirical results reveal that the EPP expanded the market share of energy-efficient digital TVs within each screen-size class. However, the EPP also induced consumers to purchase large-screen TVs. By multiplying the change in sales by the annual energy consumption, we found that the EPP increased energy consumption from TV usage.

Our empirical results sound the alarm on the application of a rebate program. When product price is used as the basis for rebates, consumers receive greater benefits when purchasing expensive products, which, in the case of TVs, are large and have many functions. In some situations, a rebate program induces consumers to purchase energy-intensive products. Therefore, the size of the rebates should be determined based on the criterion of absolute energy efficiency.

Appendix 4

A. First-stage estimation: hedonic price function

We estimated the following hedonic price function in the first stage:

$$\ln p_{tj} = \beta_0 + \beta_s \ln\left(screen_{tj}\right) + \sum_k \beta_k z_{tjk} + \omega_{tj} \tag{4.A1}$$

where $screen_{tj}$ is the average screen size in category j and z_{tjk} denotes the k's product characteristic. The result of Equation 4.A1 is reported in Table 4.A1. As presented in the table, the explanatory power of the model is quite high: the adjusted R^2 is 0.9628. Using the coefficient in Table 4.A1, we calculated the expected price of $E\left[p_j \mid Z_{mj}\right]$ and the residual μ_{tj}.

Table 4.A1 First-stage analysis: the hedonic price function (N = 750)

	Coefficient[a]	Standard Error[b]	Z Value
Constant	8.212138***	0.5222	15.73
Log of screen	1.252862***	0.052158	24.02
Period	−0.02569***	0.001703	−15.09
Days after initial sales	−0.00011**	4.78E−05	−2.26
Manufacturer dummies			
Manufacturer A	−0.31364	0.211732	−1.48
Manufacturer B	0.598298***	0.09227	6.48
Manufacturer C	0.068722	0.076279	0.9
Manufacturer D	0.122974	0.095494	1.29
Manufacturer E	−0.18918***	0.058768	−3.22
Function dummies			
LCD display dummy	−0.13315	0.960583	−0.14
Plasma display dummy	−0.23008	0.959381	−0.24
Full HD dummy	0.249683***	0.033594	7.43
DVD dummy	1.491883*	0.774552	1.93
Blu-ray dummy	−0.99313	0.79504	−1.25
HDD dummy	0.229452***	0.035044	6.55
Memory dummy	−0.55127***	0.067407	−8.18
Churner dummy	−0.28123	1.069232	−0.26
Digital pins	−0.84305***	0.190109	−4.43
HDMI dummy	0.220152***	0.029366	7.5
Modem dummy	0.209307*	0.116707	1.79
PC dummy	0.021136	0.04634	0.46
DLNA dummy	0.182375**	0.082102	2.22
acTvila dummy	−0.49123***	0.094647	−5.19
Electric program dummy	−0.20398	0.339488	−0.6
LED back light dummy	0.177482***	0.044346	4
Screen separation dummy	0.25954***	0.058526	4.43
3D dummy	0.205007	0.086351	2.37
Wald chi-square (df = 26) = 41,786.06, adjusted R^2 = 0.9628			

a *, **, and *** indicate significance at the 10%, 5%, and 1% levels.

b The standard error based on the bootstrap method was used.

Acknowledgments

An earlier version of this paper was presented at the 2013 annual meeting of the Society of Environmental Economics and Policy Studies, the 4th Congress of the East Asian Association of Environmental and Resource Economics, and 5th World Congress of Environmental and Resource Economists. We thank Hajime Katayama and Makoto Sugino for their helpful comments.

Notes

1 It may induce consumers to purchase the product more frequently.
2 Although past studies found rebound effects, they reported that the sizes of these effects were relatively small.
3 120 yen is converted into $1.
4 We follow the procedure outlined by Pakes (1994).

References

J.A. Dubin, A.K. Miedema, and R.V. Chandran. (1986) 'Price effects of energy-efficient technologies: a study of residential demand for heating and cooling', *The Rand Journal of Economics* 17, 310–25.

European Audiovisual Observatory. (2012) '22 of the 27 EU member states have implemented the 2012 analogue TV switch-off in line with European Union recommendations', Press Release. Strasbourg. Available: <http://www.obs.coe.int/about/oea/pr/mavise_2013mars_dtt_so.html> (accessed 11 March 2013).

K.S. Gallagher and E. Muehlegger. (2011) 'Giving green to get green? Incentives and consumer adoption of hybrid vehicle technology', *Journal of Environmental Economics and Management* 61, 1–15.

L.A. Greening, D.L. Greene, and C. Difiglio. (2000) 'Energy efficiency and consumption – the rebound effect – a survey', *Energy Policy* 28, 389–401.

J. Haughton and S. Sarkar. (1996) 'Gasoline tax as a corrective tax: Estimates for the United States, 1970–1991', *The Energy Journal* 17, 103–26.

J.A. Hausman. (1979) 'Individual discount rates and the purchase and utilization of energy-using durables', *The Bell Journal of Economics* 10, 33–54.

Japan Electronics and Information Technology Industries Association. (2011) 'Domestic shipment of digital terrestrial broadcast television receivers', Available: <http://www.jeita.or.jp/japanese/stat/digital/2011/index.htm> (accessed 10 June 2013). (In Japanese).

C.T. Jones. (1993) 'Another look at U.S. passenger vehicle use and the "rebound" effect from improved fuel efficiency', *The Energy Journal* 14, 99–110.

D. Khazzoom. (1980) 'Economic implications of mandated efficiency in standards for household appliances', *The Energy Journal* 1, 21–40.

K. Kim and A. Petrin. (2010) Control function corrections for omitted attributes in differentiated product models, University of Minnesota working paper. Available: <http://www.econ.umn.edu/~petrin/research.html> (accessed 8 December 2012).

Ministry of Economy, Trade and Industry. (2011) 'Policy impact of eco-point program', Available: <http://www.meti.go.jp/press/2011/06/20110614002/2011 0614002-2.pdf> (accessed 20 July 2012) (In Japanese).

Ministry of Internal Affairs and Communications. (2009) 'Survey on the prevalence rate of digital broadcasting receivers', Available: <http://www.soumu.go.jp/main_sosiki/joho_tsushin/eng/Releases/Telecommunications> (accessed 27 May 2013) (In Japanese).

Ministry of Internal Affairs and Communications. (2011) 'Japan completed analog switch off in terrestrial television broadcasting successfully', Available: <http://www.soumu.go.jp/main_sosiki/joho_tsushin/eng/Releases/Telecommunications> (accessed 20 July 2013) (In Japanese).

A. Pakes. (1995) 'Dynamic structural models, problems and prospects: mixed continuous discrete controls and market interaction', *Advances in Econometrics, Sixth World Congress*, Volume II, ed. by C. Sims, pp. 171–259, New York, Cambridge.

A. Petrin and K. Train. (2010) 'A control function approach to endogeneity in consumer choice models', *Journal of Marketing Research* 47, 3–13.

OECD. (2011) *Greening Household Behaviour: The Role of Economic Policy*, OECD, Paris.

P.M. Schwarz and T.N. Taylor. (1995) 'Cold hands, warm hearth?: Climate, net takeback, and household comfort', *The Energy Journal* 16, 41–54.

Tokyo International Communication Committee. (2010) 'Terrestrial digital broadcasting "Chideji."' Monthly Newsletter on International Exchange and Cooperation, Available: <http://www.tokyo-icc.jp/english/lespace/one/one_1003.html> (accessed 20 August 2013) (In Japanese).

5   A policy evaluation of the eco-point program

The program's impact on CO_2 reductions and the replacement of home appliances

Minoru Morita and Toshi H. Arimura

5.1. Introduction

In 2009, the eco-point program for appliances (EPP), an economic-stimulus package that promoted the purchase of energy-efficient home appliances, was launched. The goal of the program was to reduce household CO_2 emissions while also stimulating the economy. CO_2 emissions in the residential sector, and household emissions in particular, increased consistently during the first commitment period of the Kyoto protocol. Introducing the EPP thus seemed sensible, as it was aimed at promoting energy conservation by households through the use of rebates.

With this program, households could earn eco points from the government for purchasing environmentally-friendly home appliances. Points could be exchanged for gift coupons, prepaid cards, and other services or goods. Applications for points were limited to purchases of air conditioners, refrigerators, and televisions (TVs) with a rating of four or five, in accordance with the national standard for energy saving.[1] It was expected that consumers would choose to purchase energy-efficient appliances over less efficient ones that do not earn any points. This, in turn, would result in a reduction of household CO_2 emissions. In this way, the rebate program was designed to cope with the increase in household emissions.

The program did meet with some criticism, however. Some researchers and media were skeptical of the program's effectiveness. The biggest problem was the free-rider problem. Under the EPP, anyone could apply for eco points as long as they were buying a new, energy-efficient TV, air conditioner, or refrigerator. Many consumers, however, would purchase energy-efficient appliances even without such a program. In this case, the consumers were free-riding the program without changing their behavior. This is known as the free-rider problem in the subsidy for energy-efficient appliances or investment (Gillingham et al. 2009). Therefore, it is worth examining the effectiveness of the program from this perspective.

The objective of the present study is to evaluate the EPP with regard to its effectiveness in reducing household CO_2 emissions. Among the appliances eligible for eco points, we focus on refrigerators. We are primarily interested in examining refrigerators for the following reasons. First, refrigerators consume more electricity than any other household appliance, and therefore reducing their energy consumption is considered to be crucial. Second, the decision about when to use, buy, or replace refrigerators is not affected by external factors. The same does not apply for air conditioners or TVs. For instance, the amount of energy saving from an energy-efficient air conditioner depends on usage, which is heavily influenced by daily temperature. Thus, weather conditions matter. Moreover, the purchase of TVs was promoted by the introduction of digital terrestrial broadcasting. Third, refrigerators operate 24 hours a day, 365 days a year, hence emitting a more or less similar amount of CO_2 over time. This is not the case with the other appliances.

In the next section, we will provide an overview of the EPP. Section 5.3 will examine the program's impact on refrigerator sales volume. In Section 5.4, changes in sales volume will be used to calculate CO_2 reductions. Section 5.5 concludes this chapter.

5.2. The eco-point program

5.2.1. *How does the program work? How and to what extent is it effective as a policy measure?*

The EPP was aimed at promoting global warming countermeasures and stimulating the economy by encouraging the purchase of energy-efficient household appliances. The introduction of the program was also meant to diffuse TVs that support digital terrestrial broadcasting (hereafter, referred to as "DTB TVs"). The program was implemented on 15 May 2009 and continued until 31 March 2011. The government issued eco points to households that purchased energy-efficient electronic appliances. Earned points were exchangeable for other goods and service. Approximately 45 million rebate applications were submitted, and points valuing 650 billion yen were issued. Originally, one could apply for points for buying DTB TVs, air conditioners, and refrigerators that were rated four or five stars in accordance with the energy efficiency labeling system.[2] Then, starting from 1 January 2011, those eligible for application were limited to purchasing five-star DTB TVs, air conditioners, and refrigerators. It was expected that households were likely to choose energy-efficient products that were eligible for points over those that were not, and this, in turn, would reduce household CO_2 emissions.

Let us look at the design of the program. The number of points one consumer could receive from the purchase of one unit depended on the size of the refrigerator. As shown in Table 5.1, a consumer could receive 10,000 points if the size of the refrigerator was 501 liters or greater. In contrast, the amount of points was only 3,000 if the size of the refrigerator was less than 250 liters. The number of applications received was more than was initially expected by the government. Facing the risk of the shortage of the subsidy, the government

Table 5.1 Eco-point program for refrigerators

Capacity (liters)	Points per Unit	
	Before	*After the Change*
	(15 May 2009–30 November 2010)	*(1 December 2010–31 March 2011)*
501 or greater	10,000	5,000
401 of greater	9,000	5,000
201 or greater	6,000	3,000
Less than 250	3,000	2,000

redesigned the program. Namely, the size of the point given to each purchase was reduced to almost half that of the initial program after 1 December 2010.

The Ministry of the Environment (MOE), the Ministry of Economy, Trade and Industry (METI), and the Ministry of Internal Affairs and Communications (MIAC) (MOE/METI/MIAC, 2011) issued their evaluation of the program.[3] According to their assessment, the share of four- or five-star products had increased dramatically since the program was implemented: 99 percent of all DTB TVs shipped from April to December of 2010 were four- or five-star products, and 98 percent of all refrigerators and air conditioners shipped in the same period were four- or five-star products. They also reported that the effect of the program on CO_2 reductions was estimated to be 2.7 million t-CO_2/year, of which 0.65 million t-CO_2/year were from TVs, 1.29 million t-CO_2/year were from refrigerators, and 0.79 million t-CO_2/year were from air conditioners. In addition, the overall economic impact of the program was estimated to be 5 trillion yen.

One may argue, however, that the above evaluation may have overestimated the program's outcomes. The estimation does not take into account the fact that some households would have replaced their TVs, refrigerators, and/or air conditioners even if the program was not implemented. In other words, for some households, the decision to buy new TVs, refrigerators, and/or air conditioners was not necessarily affected by the incentive program. These households must be excluded when one estimates the program's impact on CO_2 reductions. In fact, Board of Audit of Japan (BOA 2012) issued an independent review of the program and criticized the program's effectiveness.

5.2.2. Alternative approach to assessing the program's outcomes

To estimate the program's exact effect, this study will take a different approach than the one used in the government's assessments. Specifically, we estimate the demand function for refrigerators using information from before the implementation of the EPP. Then, with the estimated demand function, we predict the sales of the refrigerators for the period when the EPP was implemented. By comparing the prediction to the real sales under the EPP, we can estimate the increase of sales in the refrigerators that can be purely attributed to the EPP.

The Japanese government implemented the Top Runner program for appliances (Arimura 2015), which promotes the energy efficiency of products. This means that new appliances are typically more energy efficient than older ones. Thus, consumers can reduce CO_2 emissions from appliances through the purchase of new, energy-efficient appliances. Basically, the EPP accelerated this process. With the use of the estimated demand function, we can identify how much the program accelerated the replacement of refrigerators. In this way, we can exclude free riders from the estimation of CO_2 reduction.

5.3. Empirical analysis of refrigerator sales

5.3.1. The program's impact on sales volume

Our evaluation of the EPP has two parts. First, we examine whether and to what extent the program influenced the sales volume of refrigerators. Specifically, we will identify the amount of increases in sales volume contributed by the program. Second, we will estimate the amount of CO_2 reductions achieved by the program. While MOE, METI, and MIAC might have overestimated the program's effect on CO_2 reductions, we attempt to estimate the program's pure effect by using the survival rate of refrigerators.

5.3.2. Data

Our data on refrigerators are drawn from the *Yearbook of Machinery Statistics* provided by METI (2011). Figure 5.1 shows the transition in sales volumes from January 1990 to March 2011. As shown in the figure, sales volume fluctuates every month in each year due to seasonal effects. Nevertheless, one can observe several trends in sales volume. First, sale volumes increased from 1990 to 1997. It should be noted that there was an increase in the consumption tax from 3 to 5 percent in 1997, which was followed by the Yamaichi shock[4] and led to the recession. Once again, sales peaked around 2000. After that, sales generally declined over the years. Once the EPP started (May 2009 – March 2011), however, the sales volumes started to increase. This increase seems to be attributable to the introduction of the EPP. The exact impact of the program, however, is not clear until we conduct a rigorous quantitative analysis. We will do so by estimating the demand equation for refrigerators using data from before the program was implemented.

During the expansion of the sales under the EPP, how many eco points were used? As Table 5.2 shows, more than 2 million refrigerators were eligible in 2009 and 23.7 billion points were issued. In 2010, more refrigerators were

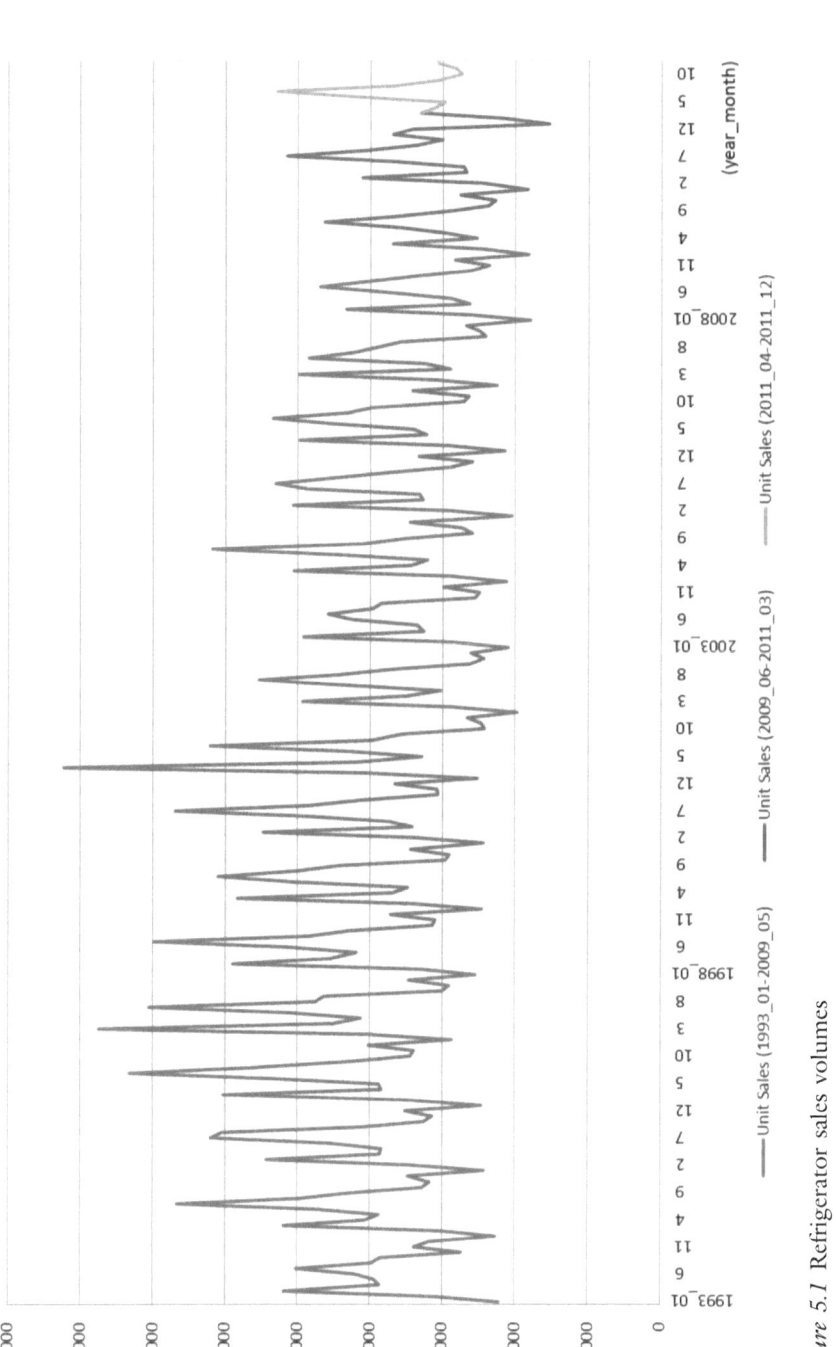

Figure 5.1 Refrigerator sales volumes
Source: METI (2011).

Table 5.2 Application of eco points for refrigerators

Fiscal Year	Points Issued	Number of Eligible Refrigerators	Share of Eco-Point Refrigerators
2009	23,779,635,000	2,143,246	63.8%
2010	32,089,779,000	3,070,267	79.2%
Total	55,869,414,000	5,213,513	

eligible for the EPP. More than 3 million units were sold in association with the program and, consequently, the points also increased by almost 50 percent. It is worth mentioning that the market share of eco-point refrigerators increased from 63.8 percent in 2009 to 79.2 percent in 2010. One can expect that producers shifted their production toward more energy-efficient types of refrigerators once the program started.

5.3.3. Demand function: data and estimation results

To identify the exact effects of the EPP on sales volumes, we estimate the demand function of refrigerators. EPP was introduced in May 2009 and terminated in March 2011. Consequently, for an accurate analysis of the program, we must estimate the demand function based on monthly data.

To estimate the demand function accurately, we will look into the details of the monthly fluctuation in sales volumes. Figure 5.2 shows the sales of the refrigerators from January to December in 2004 by month. It exhibits a typical fluctuation in refrigerator sales in the following sense. First, there is a hike in refrigerator sales in March. This is because the Japanese fiscal year and school year start in April. Therefore, a number of people move in March, and thus buy a new refrigerator for the new apartment or housing.[5] Second, there is another peak in July. This is the month when Japanese workers receive an extra payment, or what they call a "bonus" in Japanese society. Every year, we observe that Japanese households purchase new appliances in July, and the refrigerator is no exception. From these observations, we include *season dummy1 (March)* and *season dummy2 (July)* in our estimation.

We estimate the following demand function:

$$
\begin{aligned}
Sales_t = {} & \beta_0 + \beta_1 \cdot Price_t + \beta_2 \cdot Income_t + \beta_3 \cdot Sales_{t-1} \\
& + \beta_4 \cdot Electricity\_Bills_{t-1} + \beta_5 \cdot CDD\_HDD_t \\
& + \theta_1 \cdot Time\_Trend + \theta_2 \cdot Season\_Dummy1 \\
& + \theta_3 \cdot Season\_Dummy2 + \varepsilon_t
\end{aligned}
\tag{5.1}
$$

where β and θ are the parameters to be estimated and ε is the error term.

Various other factors may influence the demand for refrigerators. First, it is a function of *price*. We use the average price of refrigerators. Specifically, we calculate the price by dividing the total sales value by the total volume. This will give us the average price of refrigerators.

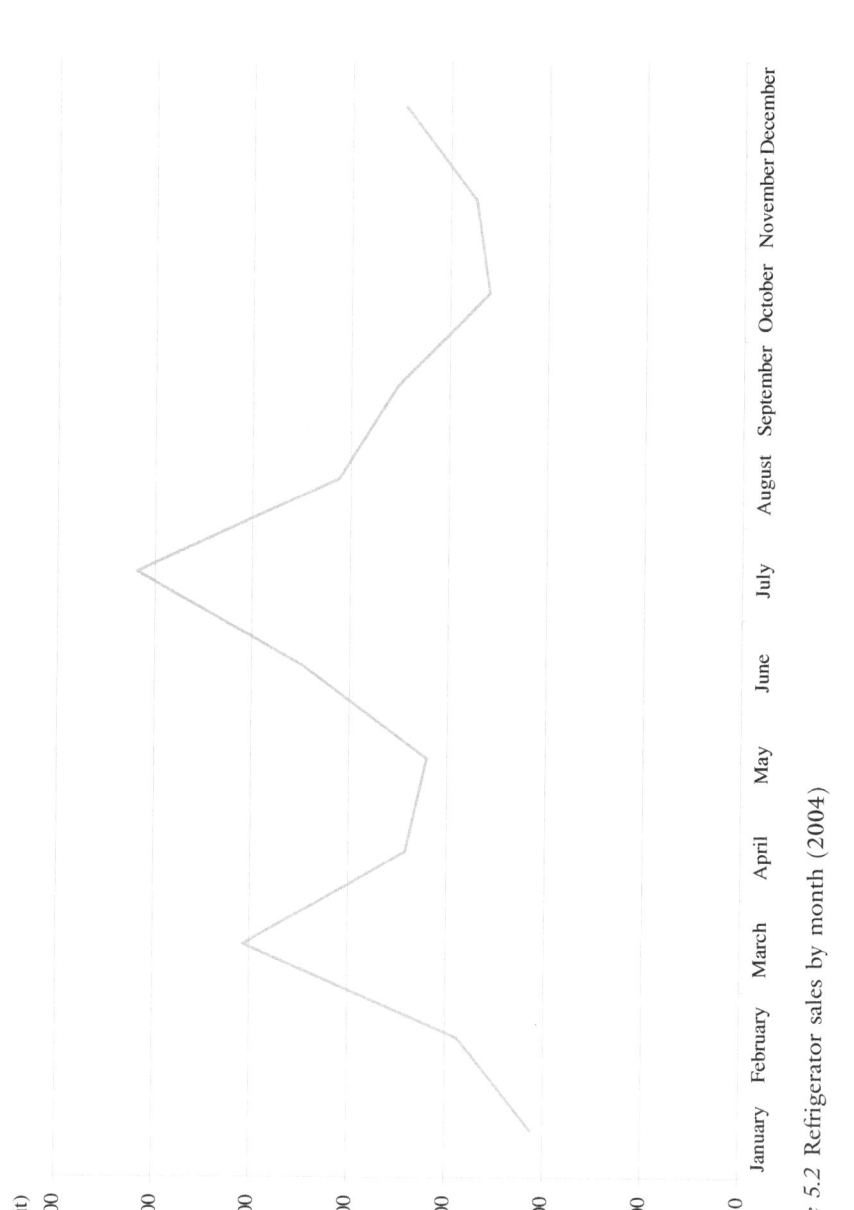

Figure 5.2 Refrigerator sales by month (2004)

Demand for refrigerators also depends on other economic variables. For example, it is expected to be influenced by the larger economic situation. If the economy is booming, consumers are more likely to purchase refrigerators. We measure the economic condition with *disposal income per household*. If the disposal income becomes smaller, consumers are less likely to purchase a refrigerator.

Temperature may also affect the purchase of refrigerators. In hot summers, consumers may be more likely to replace (or purchase) refrigerators to avoid malfunction. We use the sum of heating degree days and cooling degree days and denote *CDD_HDD*.

Consumers may have an incentive to purchase a new refrigerator if their electricity bills are high, because a new one can reduce the bill. We include *electricity bill* in the previous month to capture this aspect.

Demand for refrigerators is affected by seasonal factors, as hinted at in Figure 5.2. Thus, we include *season dummy1 (March)* and *season dummy2 (July)*, which capture the demand from movers and the demand from "bonus," respectively. Moreover, we include a variable *trend* to capture time trends.

It should be noted that the price may be endogenous in the sense that the error term may be correlated with the equilibrium price. To deal with the endogeneity issue of the price, we apply the two-stage least square. More specifically, we will use the *exchange rate* as the instrument. The inputs of refrigerators are imported, and thus the cost of the refrigerators depends on the exchange rate. Therefore, we estimate the following first-stage equation:

$$
\begin{aligned}
Price_t = {} & \alpha_0 + \alpha_1 \cdot Income_t + \alpha_2 \cdot Sales_{t-1} + \alpha_3 \cdot Electricity\_Bills_{t-1} \\
& + \alpha_4 \cdot CDD\_HDD_t + \gamma_1 \cdot Time\_Trend + \gamma_2 \cdot Season\_Dummy1 \quad (5.2) \\
& + \gamma_3 \cdot Season\_Dummy2 + \delta_1 \cdot Exchange\_Rate_t + \mu_t
\end{aligned}
$$

where α and γ are the parameters to be estimated and μ is the error term.

Table 5.3 exhibits the summary statistics of the variables used in the estimations. The EPP was introduced in May 2009 and modified in December 2010. Therefore, an accurate estimate of the impact requires the use of monthly data. For the estimation, we use the observation before the EPP was introduced. To capture the declining trend after 2000, we use the observation from January 2000 to April 2009, which results in 112 observations.

Sales volumes of refrigerators are taken from the *Yearbook of Machinery Statistics* published by METI (2011). The *price* of refrigerators is calculated by dividing total sales by sales volumes. These are also taken from the *Yearbook of Machinery Statistics*. *Income* and *Electricity Bills in the Previous Month* are taken from the *Family Income and Expenditure Survey* published by MIAC (2012). *Cooling and Heating Degree Days* is the sum of days with a temperature of 12 degrees Celsius or cooler and days 30 degrees Celsius or warmer. This data was obtained from the Japan Metrological Agency (2014). The *Index of Real Exchange Rate* is the index that was normalized to 100 in 2010 and is published by Bank of Japan.

Table 5.3 Summary statistics

Variable	Obs	Mean	Std. Dev.	Min	Max
Sales	112	357,393	112,128	179,936	822,949
Price	112	100,749	10,082	73,435	122,471
Income	112	446,520	134,786	327,139	924,744
Sales of the Previous Month	112	358,205	111,692	179,936	822,949
Electricity Bills in the Previous Month	112	9,839	8,993	6,950	103,112
Cooling and Heating Degree Days	112	11.31	12.41	0	31
Season Dummy 1 (March)	112	0.089	0.29	0	1
Season Dummy 2 (July)	112	0.080	0.27	0	1
Time Trend	112	56.50	32.48	1	112
Index of Real Exchange Rate	112	101.73	12.79	79	128

Table 5.4 presents the estimation results. The left-hand side of the table (Eq. 5.1) shows the estimation results of the first-stage regression. One can confirm that *exchange rate* is positive and statistically significant at the 1% level. This is likely to show that the cost of production will rise if the index of real exchange rate increases because the inputs of refrigerator production are imported.

The results of the demand equation estimation from the second-stage regression are shown on the right-hand side (Eq. 5.2) of Table 5.4. The coefficient of *price* is negative, as expected, although it is not statistically significant. The coefficient of *income* is positive and statistically significant at the 1 percent level. Thus, consumers are more likely to purchase a new refrigerator when their income is higher (i.e. when economic conditions are booming).

Season is an important factor in the purchase of new refrigerators. The coefficients of *season dummy1* and *season dummy2* are positive and statistically significant. This confirms that consumers are more likely to purchase new refrigerators in March and July, relative to other seasons. In Japan, because the new life begins in April, many people purchase new appliances in March.

Further, it is confirmed that weather is an important variable in the refrigerator purchase decision. In order to cool the inside of refrigerators, it is necessary to perform heat exchange. Refrigerators are more likely to break down in the summer, because the efficiency of heat radiation is reduced in hot weather. The coefficient of *CDD_HDD* is positive and statistically significant. This result suggests that consumers are likely to purchase new refrigerators to replace those that break down in hot summer weather.

Table 5.4 Estimation results

Dependent Variable	Eq. 1 (First-stage regression)		Eq. 2 (Second-stage regression)	
	Price		Sales	
	Coef.	Std. Err.	Coef.	Std. Err.
Price	–		–0.732	(0.85)
Income	0.010	(0.01)*	0.234	(0.04)***
Sales of the Previous Month	–0.006	(0.01)	0.221	(0.06)***
Electricity Bills in the Previous Month	–0.001	(0.08)	–0.255	(0.15)*
Cooling and Heating Degree Days (CDD_HDD)	–391.4	(70.19)***	–2,953	(604.78)***
Season Dummy 1 (March)	1,876	(2,516.44)	252,867	(31,957.36)***
Season Dummy 2 (July)	307.2	(2,623.23)	163,291	(26,205.32)***
Time Trend	312.9	(39.94)***	–568.5	(195.91)***
Exchange Rate	930.2	(100.39)***	–	
Constant Term	–9,609	(12,969.32)	279,714	(89,297.99)***
Adjusted R2	0.50		0.72	
Number of Samples	112		112	

Note: Asterisks ***, **, and * denote significance levels of less than 1%, 5%, and 10%, respectively.

The coefficient of *time trend* is negative and statistically significant. This finding suggests that the market for refrigerators was shrinking over time. Thus, the EPP was expected to reverse or at least stop this trend.

Using this equation, we first obtain the production volume of each month (*t*) before the program's implementation. Based on the estimation, we predict sales volumes for the case when the program was not implemented. By subtracting the predicted value from real sales, we can identify the increase in sales volumes resulting from the introduction of the program. This approach enables us to identify the increase in the number of refrigerators that have been replaced purely because of the program, while excluding those that have been replaced because they stopped functioning properly as the end of their life expectancy was nearing.

5.4. The program's impact on CO_2 reductions

In the MOE/METI/MIAC (2011) estimation, the program's impact on CO_2 reductions is calculated based on the number of applications submitted to earn eco points. As mentioned earlier, this estimation includes the case where refrigerators were replaced not because of the incentive program, but rather for other reasons, such as a product's malfunction or natural retirement. This might have led to overestimating the true amount of CO_2 reductions contributed by the program. Below, we estimate the program's net effect on CO_2 reductions by using the results we have obtained in the previous section (i.e. the increase in the number of refrigerators that have been replaced).

5.4.1. Illustration of the emission reduction

The CO_2 emission intensity for new and old refrigerators can be calculated by using "*Shinkyusan*," an online simulation service created and managed by MOE (2014). This service provides estimates with regard to replacing home appliances currently in use with more energy efficient ones, including the amount of CO_2 emissions and energy consumption that can be reduced by the replacement.

   Figure 5.3 illustrates the emission reduction effects of the EPP from a refrigerator. The emission reduction from the replacement is the difference between emissions in the absence of the program (i.e. the baseline) and emissions with the program.

   We use e^N (kWh/year) to denote the CO_2 emissions from the newly purchased refrigerator. Similarly, we use e^Y to denote the volume of CO_2 emissions from the old refrigerator purchased in year Y. Due to the improvement in the efficiency of refrigerators, we have $e^N < e^Y$. Figure 5.3 illustrates the case of refrigerators sold

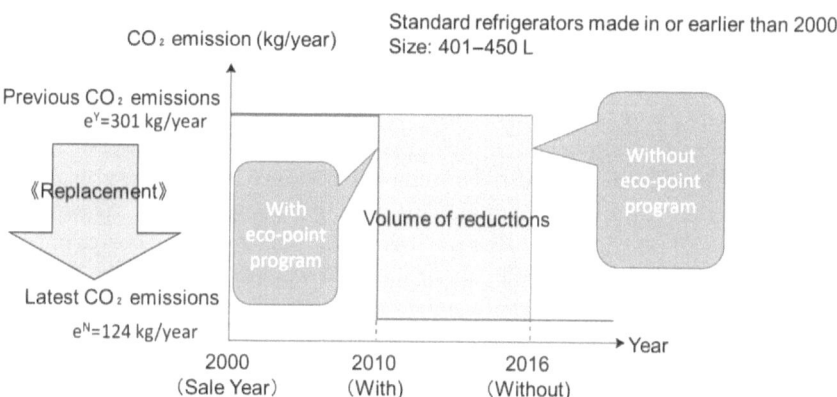

Figure 5.3 Emission reduction from refrigerator replacement by the eco-point program

in 2000. According to *Shinkyusan*, the annual CO_2 emissions from refrigerators sold in 2000 was 301 kg,[6] on average, for the size of 401 to 450 liters, whereas it is 124 kg for new ones sold in 2010. The shaded part in Figure 5.3 depicts the emissions reduction volume achieved by the EPP for the refrigerators.

It should be noted that even without the program the old refrigerators sold in 2000 would have been replaced with new ones complying with the latest energy efficiency guidelines (e^Y = 124 kg/year). Most refrigerators are replaced after 16 years.[7] Hence, even in the absence of the EPP, CO_2 emissions decrease due to technological progress. However, this type of emission reduction must be distinguished from the emission reduction achieved by the EPP. With the EPP, the same refrigerators would be replaced in earlier years. This earlier replacement led to further emission reduction, which is the effect of the EPP that we have attempted to measure.

5.4.2. Methodology: survival rate

To estimate the amount of CO_2 reduction resulting from the program, we must first figure out which model years of refrigerators were replaced with new ones. For this purpose, we will follow the approach by Iwata and Arimura (2009). This approach requires determination of the survival rate (i.e. what proportion of refrigerators are still in use among those manufactured in a particular year) of refrigerators. We take the survival rate from Tasaki (2006). Tasaki (2006) calculated the length of use of refrigerators based on their survival rate, which was identified from information on refrigerator ownership. We use Tasaki's data on refrigerators' survival rate as well as his findings on the relationship between refrigerators' life expectancy and their survival rate.

There are a few assumptions that we have to make in the estimation. First, we assume that the increase in the number of refrigerators that have been replaced corresponds to the increase in sales volumes obtained above in Section 5.3.2. Second, we assume that refrigerators are to be replaced from oldest to newest according to their year of manufacture.

In estimating the emission reduction, we must find out the transition in the number of refrigerators over time. First, we consider the case without the EPP. Let N_t^Y denote the number of refrigerators that were purchased in year Y and that are still used in the period t. This number becomes smaller every year due to natural retirement or malfunction. This transition is captured by the survival rate. Table 5A.1 in the appendix shows N_t^Y for each t and Y. For example, among refrigerators sold in 1990, about 332,000 were estimated to be used in 2009 without the program. This number was reduced to about 256,000 due to the natural replacement.

The volume of CO_2 emissions without the EPP can be obtained by combining the volume of emissions from refrigerators that have and have not been replaced. Let $E_{t,WO}^Y$ denote the emission at year t from refrigerators sold in year Y without the program. This can be constructed as follows:

$$E_{t,WO}^Y = N_t^Y \times e^Y + (N_{t-1}^Y - N_t^Y) \times e^N \qquad (5.3)$$

The first term on the right-hand side of the equation measures CO_2 generated from the old refrigerators sold in the year Y while the second term measures that from the new ones replaced in the period $t - 1$.

Next, we will calculate CO_2 emissions with the EPP. Let $\widehat{N}_t^Y$ be the number of refrigerators that were sold in year Y and that are still used in the period t after the program implementation. Let $E_{t,W}^Y$ denote CO_2 emissions at year t from the refrigerators purchased in year Y with the program implementation. We can calculate this as follows:

$$E_{t,w}^Y = \widehat{N}_t^Y \times e^Y + (\widehat{N}_{t-1}^Y - \widehat{N}_t^Y) \times e^N \tag{5.4}$$

In this calculation, we must estimate $\widehat{N}_t^Y$. This process can be done from the estimation of the demand function in the previous sections and the survival rate. We assume that the older refrigerators were replaced. Table 5A.2 in the appendix illustrates the estimated numbers. In contrast to Table 5A.1, the retirement of old refrigerators was accelerated due to the program. In 2009, the number of refrigerators sold in 1989 is estimated to be only 3,790 with the program, while it is 252,806 without the program. That is, most of refrigerators sold in 1989 had been replaced before 2010 by the EPP.

From the equations above, we can estimate the net impact of the reduction in CO_2 emissions from the program, ER, as follows:

$$ER = \sum_t \sum_Y \left(E_{t,WO}^Y - E_{t,W}^Y \right) \tag{5.5}$$

5.4.3. Emission reduction

Based on the estimation results (i.e. the increase in sales volumes due to the implementation of the program) and the survival rate which was calculated from the sales volumes between 1985 and 2011, below we identify the manufacturing years of refrigerators that are expected to be replaced. The results are presented in Table 5.5. As shown in the table, the program appears to have promoted the replacement of refrigerators that were made in or earlier than 1994.

Using *Shinkyusan*, we then calculate the amount of CO_2 reductions achieved if refrigerators produced in or earlier than 1994 are replaced with new ones between 2009 and 2011. In our estimation of CO_2 reduction, we used the CO_2 emissions from the refrigerators with the capacity of 400 to 450 liters. The CO_2

Table 5.5 Manufacturing years of refrigerators (RF) to be replaced

Years	Sales Volumes to be Replaced	Manufacturing Years
2009	622,092	Made in 1986–1989
2010	1,358,761	Made in 1989–1993
2011	142,122	Made in 1993–1994

Table 5.6 CO_2 reductions for replacing refrigerators made in or earlier than 1994

Size of Refrigerators	Amount of Reductions (kg-CO_2/year)	
	Upper Bound	Lower Bound
401–450 L	350	450

emission from old refrigerators cannot be pinned down exactly, because we cannot identify the information of each replaced refrigerator. For example, according to *Shinkyusan*, CO_2 emissions from refrigerators sold in 1994 ranged from 512 to 569 kg per year. Although the difference depends on the particular make and other characteristics, this information is not available for our analysis. Thus, we use upper bound and lower bound in estimating the emission reduction. Table 5.6 presents the upper bound and the lower bound of the emission reductions if refrigerators produced in or earlier than 1994 are replaced with those made by five domestic brands eligible for eco points.[8]

5.4.3. Results

We estimate the total amount of CO_2 reductions by multiplying the number of refrigerators that have been replaced with the amount of reductions achieved by replacing a refrigerator. This is because we assume that the number of refrigerators replaced corresponds to the increase in sales volumes. Table 5.7 shows the results.

Table 5.7 illustrates the upper bound and lower bound of the estimated CO_2 reduction from the EPP for each year. Columns (1) and (2) exhibit the emission reduction during the period when the EPP was implemented. The lower bound of the estimated reduction from the program is 743,000 t-CO_2 while the upper bound is 955,000 t-CO_2. This estimated impact is smaller than those from MOE/METI/MIAC (2011) or BOA (2012). MOE/METI/MIAC (2011) estimated the reduction from the program as 1.29 million t-CO_2. The estimated reduction from BOA (2012) was 998,000 t-CO_2. The difference from

Table 5.7 The program's effect on CO_2 reductions

	CO_2 Emission Reduction from the Program (unit: 10,000 t-CO_2)					
	2009–2011		2012–2018		Total	
	Lower Bound (1)	Upper Bound (2)	Lower Bound (3)	Upper Bound (4)	Lower Bound (5)	Upper Bound (6)
2009	21.8	28.0	9.7	12.5	31.5	40.5
2010	47.6	61.1	83.3	107.1	130.9	168.3
2011	5.0	6.4	49.8	64.0	54.7	70.4
Total	74.3	95.5	142.8	183.6	217.1	279.1

our estimation arises from the treatment of the *free riders*. In our analysis, we distinguish the consumers who purchased refrigerators with the assistance from the eco points and those who purchased them even without the program. Thus, we were able to identify the net impacts of the EPP by excluding the free riders from our estimation.

On the one hand, MOE/METI/MIAC and BOA overestimated the impact of the program by ignoring the free-rider problem. On the other hand, they underestimated the impact by ignoring the impacts of the program after the EPP. Those newly purchased refrigerators are still more energy efficient than older ones. We estimated these emission reductions, as shown in columns (3) and (4). We found that the impact of the program is greater after 2012 and will last until 2018.

Overall emission reduction impacts are shown in columns (5) and (6). The total emission reduction was estimated to range from 2.17 to 2.79 million tons. Thus, MOE/METI/MIAC underestimated the overall impacts.

Based on our analysis, one can calculate the cost-effectiveness of the subsidy (i.e. how much money did the government spend to reduce 1 ton of CO_2?). Under the EPP, eco points worth of 56 billion yen were given to the purchase of refrigerators. From the estimated CO_2 reduction, one can conclude the following. The Japanese government spent somewhere between 20,062 and 25,794 yen for the reduction of 1 ton of CO_2.

5.5. Conclusions

This study has examined whether and to what extent the EPP promoted CO_2 reductions from the purchase of energy-efficient refrigerators among households. Specifically, we have considered changes resulting from the program's implementation in two respects: changes in sales volumes of refrigerators and the amount of CO_2 reductions. Our approach has enabled us to identify the environmental effects genuinely contributed by the program. This distinguishes our study from the assessment by the MOE/METI/MIAC or BOA. In their assessments, the increases in the number of refrigerators that have been replaced during the EPP implementation were assumed to correspond to the number of applications that were submitted to earn eco points. Consequently, in their assessment, CO_2 reductions contributed by the program are likely to have been overestimated.

Indeed, we found that the impact from 2009 to 2011 was much smaller than the government estimates. The estimated CO_2 reduction in MOE/METI/MIAC (2011) was approximately 1.29 million $t\text{-}CO_2$. In contrast, our studies estimated that the reduction ranged from 743,000 $t\text{-}CO_2$ to 955,000 $t\text{-}CO_2$. Thus, the reduction by the program was found to be between 58 and 74 percent of MOE/METI/MIAC's estimate. This overestimation can be attributed to the government's ignorance of the free-rider problem. In summary, we can conclude that the Japanese government overestimated the effects of the program in terms of CO_2 reduction.

Based on our estimation, we can conclude that the government spent more than 20,000 yen per 1 ton of CO_2 reduction. It should be noted that our approach is underestimating this cost in the following sense. In our estimation of emission reduction, we took the MOE/METI/MIAC approach (i.e. we compare the emission intensity of old refrigerators with new ones). This approach potentially overestimates the emission reduction relative to the BOA approach. In the latter, the difference between an eco-point-qualified refrigerator and non-qualified one was used to measure the emission reduction. Consequently, there is a possibility that our estimate of the tax expenditure per ton of CO_2 reduction is underestimated. This result has an important implication for government climate policy. If the objective of the program had been purely the reduction of CO_2, it would have been more efficient to purchase Certified Emission Reduction (CER) emission credits generated from the Clean Development Mechanism, because the government could have purchased 1 ton of CERs, one-tenth the budget of the EPP.

Obviously, the reduction of CO_2 was not the only objective of the program; the program was also meant to cope with the recession triggered by the Lehman collapse. Thus, it may not be appropriate to evaluate the program purely from the standpoint of CO_2 reduction. The Japanese government, however, introduced a carbon tax in October 2012; the revenue from the carbon tax goes to a special account, and so far use of the tax revenue is limited to CO_2 reduction measures or new technology development. The tax rate is gradually increasing, and total revenue is expected to reach 262 billion yen.[9] Thus, taxpayers should be very careful in watching how the government spends the carbon tax revenue. If the revenue is used in an efficient way, it will bring large reductions in emissions; it will be a big waste if it is not.

In addition, the dynamic impact of the program should be also evaluated. The EPP may have changed firms' behavior. Namely, facing the introduction of the program, they may have put more effort in producing more energy-efficient refrigerators. Thus, the program should be evaluated from the viewpoint of technological innovation as well (Arakawa & Akimoto 2014).

This study assumed that appliances are to be replaced from oldest to newest. Although this assumption may make sense given that each appliance has a life expectancy, what we observe in real life might be different. In the study, we did not consider variations in refrigerator capacity. Instead, we considered the average capacity of all appliances. It is possible; however, that as appliances are replaced with new ones their kilowatts, size, and capacity become greater and enhanced. In this regard, the data and conditions we used in our analysis may not have provided a thorough and accurate evaluation. Nonetheless, we hope that the study has provided perspectives and insights different from those given by MOE/METI/MIAC or BOA.

Appendix

Table 5A.1 Estimated number of survived refrigerators without the eco-point program

Sale Year	Number of Refrigerators									
	2009	2010	2011	2012	2013	2014	2015	2016	2017	2018
1986	68,487	22,829	0	0	0	0	0	0	0	0
1987	127,268	76,361	25,454	0	0	0	0	0	0	0
1988	177,322	126,659	75,995	25,332	0	0	0	0	0	0
1989	252,806	176,964	126,403	75,842	25,281	0	0	0	0	0
1990	332,440	255,723	179,006	127,862	76,717	25,572	0	0	0	0
1991	436,510	333,802	256,771	179,739	128,385	77,031	25,677	0	0	0
1992	483,788	391,638	299,488	230,375	161,263	115,188	69,113	23,038	0	0
1993	692,648	469,213	379,839	290,465	223,435	156,404	111,717	67,030	22,343	0
1994	1,077,965	759,475	514,483	416,486	318,490	244,992	171,494	122,496	73,498	24,499

Table 5A.2 Estimated number of survived refrigerators with the eco-point program

Sale Year	Number of Refrigerators									
	2009	2010	2011	2012	2013	2014	2015	2016	2017	2018
1986	0	0	0	0	0	0	0	0	0	0
1987	0	0	0	0	0	0	0	0	0	0
1988	0	0	0	0	0	0	0	0	0	0
1989	3,790	0	0	0	0	0	0	0	0	0
1990	332,440	0	0	0	0	0	0	0	0	0
1991	436,510	0	0	0	0	0	0	0	0	0
1992	483,788	0	0	0	0	0	0	0	0	0
1993	692,648	94,268	0	0	0	0	0	0	0	0
1994	1,077,965	759,475	448,673	363,212	277,750	213,654	149,558	106,827	64,096	21,365

Acknowledgment

Toshi Arimura appreciates the financial support by JSPS KAKENHI Grant Number 24530267. Both authors are grateful to the financial aid from the Center for Global Partnership, Japan Foundation.

Notes

1 The highest ranking is five stars.
2 This system was adopted uniformly across the nation. The system mandated retailers of home appliances to display the energy efficiency label on their products. The label provides information about a product's energy efficiency performance, including the product's energy efficiency achievement rating, with one star being the lowest and five stars being the highest, as well as an estimate of annual electricity charge for the product.
3 See METI (2011) for further details. The document was prepared by the Ministry of the Environment, the Ministry of Economy, Trade and Industry, and the Ministry of Internal Affairs and Communications.
4 Yamaichi Securities was one of the four largest securities trading companies in Japan. The company collapsed after the uncovering of a fraud. The collapse led to a serious recession of the Japanese economy.
5 Japanese apartments and rental housing do not have refrigerators in place. Therefore, people must bring their own refrigerator into the apartment/housing when they move in.
6 The CO_2 emissions from a refrigerator sold in 2000 range from 284 kg to 318 kg per year. The difference depends on the make and other characteristics. Since we cannot identify the information of each replaced refrigerator, we will use upper bound and lower bound in estimating the emission reduction.
7 This is obtained from the survival rate from Tasaki (2006).
8 The refrigerators considered here are those produced by five domestic brands.
9 http://www.env.go.jp/policy/tax/about.html.

References

Arakawa, J., and Akimoto, K. (2014) 'Assessments of Japanese energy efficiency policy measures in residential sector', *Journal of the Japan Institute of Energy* 94, 333–9.

Arimura, T.H. (2015) 'Japanese domestic environmental policy with a focus on climate change and air pollution policy' in S.Managi (Eds.) The Routledge *Handbook of Environmental Economics in Asia*, 516–531 Oxon, Routledge.

Board of Audit of Japan. (2012) 'Effectiveness of subsidy program for green appliances promotion', Available: <http://www.jbaudit.go.jp/pr/kensa/result/24/pdf/241011_zenbun_1.pdf> (accessed 1 October 2014) (In Japanese).

Gillingham, K., Newell, R., and Palmer, K. (2009) 'Energy efficiency economics and policy', *Annual Review of Resource Economics* 1, 597–620.

Iwata, K., and Arimura, T.H. (2009) 'Economic analysis of a Japanese air pollution regulation: An optimal retirement problem under vehicle type regulation in the NOx–Particulate Matter Law', *Transportation Research Part D* 14, 157–167.

Japan Metrological Agency. (2014) *Weather Observation Data*, Available: <http://www.jma.go.jp/jma/menu/menureport.html> (accessed 6 March 2014) (In Japanese).

Ministry of Economy, Trade and Industry. (2011) *Yearbook of Machinery Statistics* (revised report), Available: <http://www.meti.go.jp/statistics/tyo/seidou/result/ichiran/08_seidou.html> (accessed 6 March 2014) (In Japanese).

Ministry of the Environment. (2014) *Shoene-Seihin Navigation Shinkyusan*. Available: <http://shinkyusan.com/index.html#/index/top> (accessed 6 March 2014) (In Japanese).

Ministry of the Environment, Ministry of Economy, Trade and Industry and Ministry of Internal Affairs and Communications. (2011) *Effects of the Home Appliance Eco-Point System Policy*, Available: <http://www.meti.go.jp/press/2011/06/201 10614002/20110614002-2.pdf> (accessed 1 October 2014) (In Japanese).

Ministry of Internal Affairs and Communications (2012) *Family Income and Expenditure Survey*. <http://www.stat.go.jp/english/data/kakei/>.

Tasaki, T. (2006) 'An evaluation of actual effectiveness of the recycling law for electrical home appliances', *Research Report from the National Institute for Environmental Studies* no. 191. (In Japanese).

6 Did the purchase subsidy for energy-efficient appliances ease electricity shortages after Fukushima?

Kenichi Mizobuchi and Kenji Takeuchi

6.1. Introduction

Emissions of carbon dioxide from Japanese households increased 59.7 percent between 1990 and 2012, a far higher rate than that from the industrial sector (−13.4 percent) and the transport sector (4.1 percent) (Figure 6.1), primarily because of increasing household electricity demand. Figure 6.2 shows that approximately 47.6 percent of Japanese household energy consumption in 2012 was from electricity, followed by oil (21.4 percent), and town gas (18.8 percent). Electricity consumption by Japanese households has increased 56.5 percent since 1990, while oil consumption has increased 1.8 percent and town gas consumption has increased 26.2 percent.

To stimulate the purchase of eco-friendly home electronic appliances and reduce household electricity consumption, the Japanese government implemented the "eco-point program for appliances" (hereafter, EPP) from May 2009 to March 2011. The program was aimed at reducing emissions of carbon dioxide as well as lifting the economy out of its decade-long slump. It issued "eco points" to those who purchased energy-saving home electronic appliances[1] that could then be exchanged for rewards such as transit passes, electronic money, gift certificates, and local products. The Japanese Ministry of Economy, Trade and Industry (METI), the Ministry of the Environment (MOE), and the Ministry of Internal Affairs and Communications (MIAC) (2010) evaluated the EPP and found that during its implementation period it reduced carbon dioxide emissions by 2.7 million tons. While the policy evaluation suggests that there was a considerable reduction of greenhouse gases, it was based on the theoretically expected improvement in energy efficiency of equipment, and not on the actual savings or behavior of the households that purchased the equipment. Therefore, although the EPP stimulated the purchase of energy-efficient equipment,[2] it is unclear if it led to an actual reduction in electricity use. To evaluate the policy more rigorously, the evaluation must be based on observed behavior.

This study analyzes actual energy savings by Japanese households, focusing on air conditioner purchases under the EPP. For the empirical analysis, we conducted a questionnaire survey in the summer of 2012 and asked households in Osaka and the Matsuyama area whether they had purchased energy-efficient air conditioners recommended by the EPP, their level of usage of air conditioners,

Figure 6.1 Increasing rate of CO_2 emissions (from 1990 to 2012)
Data source: Greenhouse Gas Inventory Office of Japan (2014).

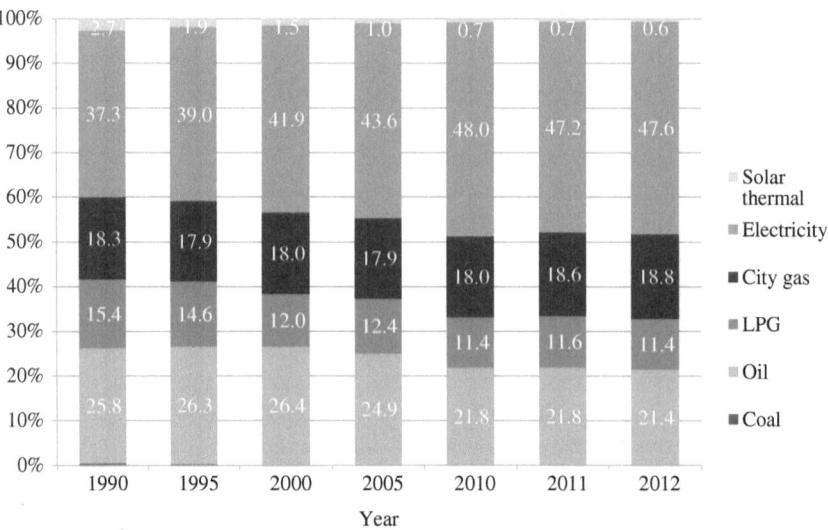

Figure 6.2 Household energy consumption according to energy source
Data source: Energy Data and Modelling Center (2014).

as well as their socioeconomic and other information. We also asked the participants to provide their electricity bill for August 2012. By combining the above information, we can empirically investigate to what extent the program contributed to the power-saving behavior of the households.

Understanding households' electricity-saving behavior provides an important implication for Japanese society after the Fukushima Dai-ichi (or Fukushima 1) nuclear disaster. After the Great East Japan Earthquake in March 2011 that triggered the disaster,[3] nuclear power plants in Japan were closed due to safety concerns, which caused power shortages in the subsequent summer and winter, when electricity demand was highest. Therefore, after 2011, the Japanese government requested power saving across all sectors as well as among households (Nishio and Ofuji 2013). During the target period of our analysis, in the summer of 2012, the power-saving request was implemented in various regions in Japan at different levels of reduction (see Section 6.3 for more details). Since this request could have an impact on the power-saving behavior of households, we considered it when analyzing the energy-saving effect of the purchase of eco-friendly air conditioners. This study targeted two regions that had different levels of power-saving request: one with a high level (Osaka) and another with a low level (Matsuyama).

The rest of this chapter is organized as follows: Section 6.2 describes the background of the study. It explains why and how the EPP was introduced and the electricity deficit after the Great East Japan Earthquake of March 2011. Section 6.3 discusses previous studies on household energy-saving behavior. Section 6.4 describes the data and the method of the empirical analysis and discusses the results we obtained. In particular, we analyzed households' purchase of energy-efficient air conditioners and its effects on their level of power saving. Section 6.5 presents the conclusions of the study and suggestions for future studies.

6.2. Background

In 2009, the Japanese government started the EPP. The goal of the EPP was to promote the development of major environmental and energy conservation technologies, such as solar photovoltaic technology, fuel-efficient cars, and eco-friendly home electronics. The Japanese government allocated about 690 billion yen ($1.00 = 101.6 yen on 15 July 2014) to MOE, METI, and MIAC for the project. These three ministries developed the EPP for home electronics, which awarded eco points to purchasers of energy-saving appliances (or "green" home appliances). By promoting the use of energy-saving home appliances, the government sought to mitigate climate change and stimulate economic growth. The energy-saving home appliances covered in the project were air conditioners, refrigerators, and terrestrial digital TVs. The program was in effect from 15 May 2009 to 31 March 2011.

The total number of eco points awarded for each appliance was about 5 percent of the purchase price for air conditioners and refrigerators and about 10 percent of the purchase price for terrestrial digital TVs (See Table II.4 on page 73). In general, these points could be exchanged for various rewards, where 1 point = 1 yen. For example, if a household purchased an air conditioner with a power consumption of 3.6 kW, it received 9,000 points in the form of a receipt. The purchaser then filled out a reward application form and selected the preferred reward (e.g. transit passes, electronic money, gift certificates, and local products) and mailed it with the product warranty and sales receipt to the secretariat of the EPP. Within one or two months, the purchaser received the reward.

Table 6.1 shows the unit sales and total number of eco points awarded for the three product types covered in the EPP. It shows that terrestrial digital TVs had the most unit sales and total number of eco points awarded among the three product types, accounting for 72 percent of unit sales and about 84 percent of the total number of points. In contrast, air conditioners accounted for only about 16 percent of unit sales and about 9 percent of the total number of points, while refrigerators accounted for only about 12 percent of unit sales and about 7 percent of the total number of points. The reason why terrestrial

Table 6.1 Unit sales and total eco points awarded to home electric appliance categories

	Unit Sales (15 May 2009– 30 November 2010)	Unit Sales (1 December 2010– 31 March 2011)	Total	Rate (percentage)
Air conditioners	7,308,627	70,953	7,379,580	16.10
Cooling capacity				
Under 2.2 kW	2,603,361	10,816	2,614,177	5.70
2.5kW, 2.8 kW	2,699,508	32,156	2,731,664	5.96
Over 3.6 kW	2,005,758	27,981	2,033,739	4.44
Refrigerators	5,086,849	171,328	5,258,177	11.50
Volume				
Under 250 L	982,602	581	983,183	2.14
251–400 L	1,147,029	3,311	1,150,340	2.51
401–500 L	1,857,818	108,073	1,965,891	4.29
Over 501 L	1,099,400	59,363	1,158,763	2.53
Terrestrial digital TVs	30,223,656	2,978,721	33,202,377	72.40
Size				
Under 26 V	6,504,594	602,257	7,106,851	15.50
26 V, 32 V	13,203,519	1,381,247	14,584,766	31.82
37 V	2,455,992	171,406	2,627,398	5.73
40 V, 42 V	6,106,725	643,570	6,750,295	14.73
Over 46 V	1,952,826	180,241	2,133,067	4.65
Total	42,619,132	3,221,002	45,840,134	100.00

	Eco Points (15 May 2009– 30 November 2010)	Eco Points (1 December 2010– 31 March 2011)	Total	Rate (percentage)
Air conditioners	52,568,544	300,977	52,869,521	9.28
Cooling capacity				
Under 2.2 kW	15,620,166	32,448	15,652,614	2.75
2.5kW, 2.8 kW	18,896,556	128,624	19,025,180	3.34
Over 3.6 kW	18,051,822	139,905	18,191,727	3.19
Refrigerators	37,544,342	848,275	38,392,617	6.74
Volume				
Under 250 L	2,947,806	1,162	2,948,968	0.52
251–400 L	6,882,174	9,933	6,892,107	1.21
401–500 L	16,720,362	540,365	17,260,727	3.03
Over 501 L	10,994,000	296,815	11,290,815	1.98
Terrestrial digital TVs	456,482,661	22,211,125	478,693,786	83.99
Size				
Under 26 V	45,532,158	2,409,028	47,941,186	8.41
26V, 32 V	158,442,228	8,287,482	166,729,710	29.25
37 V	41,751,864	1,371,248	43,123,112	7.57
40 V, 42 V	140,454,675	7,079,270	147,533,945	25.89
Over 46 V	70,301,736	3,064,097	73,365,833	12.87
Total	546,595,547	23,360,377	569,955,924	100.00

Data source: Board of Audit of Japan (2012).

digital TVs accounted for the most number of points in the program is that analogue broadcasting in Japan was replaced by terrestrial digital broadcasting on 24 July 2011. In addition, the mid-sized categories for air conditioners (2.5 kW and 2.8 kW) and refrigerators (401–500 L) and the smaller-sized category for terrestrial digital TV (less than 32 V), had the most unit sales.[4]

There is a considerable difference in the amount of the electricity consumption among the three targeted electronic appliances. The EPP targeted 2,173 air conditioners, 840 refrigerators, and 1,364 terrestrial digital TVs. Table 6.2 shows the lowest and highest power consumptions in each product category. It shows that air conditioners have the largest power consumption among all three product types. For example, the smallest air conditioner category (i.e. 2.2 kW) has a power consumption of 612 kWh, larger than the highest power consumptions for refrigerators (510 kWh for the 251–400 L category) and

Table 6.2 Highest and lowest power consumptions of home electronic appliances recommended by the eco-point program

Air conditioners		2.2 kW	2.5kW, 2.8 kW	Over 3.6 kW		
	Min	612 kWh	658 kWh	1,110 kWh		
	Max	760 kWh	1,079 kWh	3,162 kWh		
Refrigerators		Under 250 L	251–400 L	401–500 L	Over 501 L	
	Min	160 kWh	310 kWh	220 kWh	220 kWh	
	Max	450 kWh	510 kWh	450 kWh	490 kWh	
Terrestrial digital TVs		Under26 V	26–32 V	37 V	40–42 V	Over46 V
	Min	33 kWh	42 kWh	69 kWh	87 kWh	87 kWh
	Max	91 kWh	155 kWh	218 kWh	240 kWh	498 kWh

Data source: Board of Audit of Japan (2012).

terrestrial digital TVs (498 kWh for the over 46 V category). In comparing the highest power consumptions of the smallest and largest categories of the three types of home electronics, one can see that the power consumption of the smallest air conditioners is 1.68 times that of the smallest refrigerators and 8.35 times that of the smallest terrestrial digital TVs, and the power consumption of the largest air conditioners is 6.45 times that of the largest refrigerators and 6.35 times that of the largest terrestrial digital TVs. These figures suggest that although unit sales for terrestrial digital TVs were about 4.5 times higher than those for air conditioners, the impact of air conditioners on electricity saving would be the highest among the three types of appliances. Consequently, the success of the EPP might have depended greatly on households' actual usage of air conditioners.

Electricity saving was especially important for the Japanese economy after the Fukushima Dai-ichi nuclear power plant accident in March 2011. The accident led to the shutdown of nuclear power plants all over the country. The nuclear power plants had supplied about 25 percent of the total power in Japan, and their shutdown caused power shortages, especially in the summer and winter, when demand is at its peak. Thus, to balance power supply and demand, the Japanese government required all sectors to save power after the accident. In contrast to industrial sectors, it is more difficult to regulate the power usage of households. Therefore, the Japanese government encouraged households to save power by various measures, such as through the power-saving request, providing information on power-saving methods. Two studies have examined the impact of the power-saving request after Fukushima. From the data they collected through a Web-based survey of 1,500 households, Nishio and Ofuji (2013) found that power saving in the summer of 2012 was about 10 percent higher than that in the summer of 2010. Tanaka and Ida (2013) examined the voluntary power saving of households after the Fukushima Dai-ichi nuclear power plant accident by conducting a Web-based survey in the summers of 2011 and 2012 in the Kanto and Kansai regions. They found a high level of voluntary power saving among households during the survey periods as well as a difficulty in sustaining it. Arikawa et al. (2014) examined the relationship between individual electricity demand and nuclear acceptance after the Fukushima Dai-ichi nuclear disaster. The Web-based survey data, which included about 800 households, demonstrates that opponents of nuclear power used electrical appliances less intensively at home and reduced their electricity consumption during the power shortage period. On the other hand, supporters of nuclear power had high energy demands and did not reduce their electricity consumption.

The power-saving request was a government policy that required firms and households to reduce their electricity consumption voluntarily by the minimum savings required for a power company to ensure a stable supply of electricity. The policy was implemented by most of the 10 companies in the Japanese power industry,[5] which exclusively supply electricity to 10 regions (Hokkaido, Tohoku, Kanto, Hokuriku, Tokai, Kansai, Tyugoku, Shikoku, Kyusyu, and Okinawa). The levels of power saving requested varied across the 10 regions according to a power company's degree of dependence on nuclear power. The power-saving request was implemented from 1 July to 30 September in 2011 and 2012, from 9:00 to 20:00 on weekdays. Table 6.3 shows the actual levels of power saving requested of each region in 2011 and 2012, with 2010 as the base year of power saving. The table shows that in 2011, the Kanto region, whose electric power was supplied by the Fukushima nuclear power plants, was required to reduce its power consumption by 15 percent compared to 2010. Almost all nuclear power plants in Japan were shut down in 2012, and the power-saving request was implemented in most of the regions,[6] although it was

Table 6.3 Level of power saving requested by region

Region	Power Company	2012	2011
Hokkaido	Hokkaido Electric Power Co.	7%	–
Tohoku	Tohoku Electric Power Co.	–	15%
Kanto	Tokyo Electric Power Co.	–	15%
Hokuriku	Hokuriku Electric Power Co.	5%	–
Tokai	Chubu Electric Power Co.	5%	–
Kansai	Kansai Electric Power Co.	10%	10%
Tyugoku	Chugoku Electric Power Co.	3%	–
Shikoku	Shikoku Electric Power Co.	5%	–
Kyusyu	Kyusyu Electric Power Co.	10%	–
Okinawa	Okinawa Electric Power Co.	–	–

Note: The level of power saving requested indicates the reduction in power consumption requested by the government compared to 2010.

not implemented in Tohoku and Kanto in 2012, because by then nuclear power was successfully replaced by thermal power in those regions.

Our study examined households' electricity consumption in the summer of 2012, during which the demand for power exceeded the supply. We focused on two regions (Kansai and Shikoku), which were requested to reduce power consumption by 10 percent in Kansai, and by 5 percent in Shikoku, respectively.

The power consumption of households in the summer is largely dependent on their usage of air conditioners. Thus, this study used household electricity consumption data for the summer of 2012 to examine the power-saving effect of energy-efficient air conditioners. We focused on air conditioners recommended by the EPP and compared the power consumptions of purchasers of such air conditioners, purchasers of non-recommended air conditioners, and non-purchasers of new air conditioners.

The benefit of compliance to the government's power-saving request, the avoidance of sudden blackouts, can be enjoyed by everyone. In other words, by free riding, a person who does not save power still benefits from others' efforts to save power to prevent power outages. We can consider that people who do not free ride and who conscientiously save power are rather altruistic. Thus, in the implementation of the power-saving request the different levels of altruism could have affected power-saving behavior. This should also be taken into consideration when analyzing the electricity-saving behavior under the power-saving request.

6.3. Literature review

6.3.1. Energy-saving behavior

This section provides an overview of previous studies on households' energy-saving behavior. These studies can be divided into two categories according to their focus: energy-efficient investment and energy-saving behavior. Energy-efficient investment involves introducing energy-efficient equipment such as household electronic appliances, thermal insulation materials, and solar panels that enable households to reduce their energy consumption without changing their habits. In contrast, energy-saving behavior involves reducing energy consumption without introducing energy-efficient equipment. Further, the analysis of energy-saving behavior can be divided into two approaches, namely (1) the discrete-continuous approach and (2) the conditional demand approach. The former conducts a discrete choice analysis of energy equipment and analyzes energy demands based on the results (Dubin and McFadden 1984; Halvorsen and Larsen 2001; Nesbakken 2001). The latter conducts an energy demand analysis given the energy equipment (Leth-Petersen and Togeby 2001; Meier and Rehdanz 2010). In what follows, we review previous studies on households' energy saving using micro data.

Studies on energy-efficient investment have analyzed influencing factors (e.g. socioeconomic characteristics and government policies) that affect household investment decisions on equipment that uses energy. Krumm (1983) conducted a pioneering study on this topic using micro data from 1,520 American households. Estimating a multinomial logit model that distinguished between purchase of room and central air conditioners, he found that household income, housing unit characteristics, and climate conditions significantly affected investment. Cameron (1985) also used American household micro data to analyze the adoption of insulated windows and double-glazed windows. She found that cost of investment, energy price, and household income significantly influenced adoption decisions. Scott (1997) analyzed three types of investments—attic insulations, hot water cylinders, and low-energy light bulbs—using data from 1,200 Irish households. He found that household income, type of home ownership, amount of potential energy savings, and time and effort to find optimal equipment significantly affected energy-efficient investment. Using data from 305 households in Switzerland, Banfi et al. (2008) evaluated willingness to pay (WTP) for several energy-saving investments by conducting a choice experiment and showed that households placed significant value on investment benefits, such as individual energy savings benefit, environmental benefits, and comfort benefits. Grosche and Vance (2009) examined WTP in German households and revealed a significant influence of the costs and benefits of energy-saving investments. Furthermore, Nair, Gustavsson, and Mahapatra (2010) analyzed data from 3,000 households in Switzerland and found that households with higher energy consumption expenditures tended to make energy-efficient investments. A limitation of these studies, however, is that they used discrete dependent variables. To address this shortcoming, some studies began to use a Tobit model with investment expenditure as the dependent variable (Mendelsohn 1977; Montgomery 1992; Mahapatra and Gustavsson 2008; Charlier 2013). With this approach, they were able to examine not only the factors but also their impact on investment. These studies indicated that households that purchase energy-efficient equipment are generally high-income, high-energy consuming, and owner-occupier households.

Dubin and McFadden (1984) were the first to use micro data to examine households' energy-saving behavior: they surveyed 3,249 American households and analyzed household electricity demand by using a discrete-continuous approach with consideration of socioeconomic characteristics, energy prices, and ownership of electronic appliances. A number of studies have since also used micro data or micro panel data to examine energy-saving behavior across several countries (Garbacz 1985; Green 1987; Branch 1993; Jung 1993; Nesbakken 1999; Vaage 2000; Halvorsen and Larsen 2001; Wu, Lampietti, and Meyer 2004). Vaage (2000) analyzed 1,306 households in Norway and estimated the choice of energy source (electricity; electricity and wood; electricity and oil; and electricity, wood, and oil). They also estimated these sources' demand functions and revealed that household income affected energy source decisions but not the amount of energy demand. Meier and Rehdanz (2010) used panel data from 5,000 English households and analyzed the determinants of space-heating expenditures. They investigated the effects of socioeconomic characteristics, housing type, energy prices, and other variables on energy-saving behavior and found that homeowners tended to use more energy than those who rented, likely because of the poor energy efficiency of detached houses (homeowners) compared to apartments (renters).

A limited number of studies have considered the effect of households' past energy-efficient investments on energy demand. Halvorsen and Larsen (2001) estimated the short- and long-term price elasticity of electricity using Norwegian household panel data to analyze the influence of electronic appliance ownership on power demand. They found that the electronic appliances affected households' power consumption significantly. Using household panel data for the Netherlands, Berkhout, Ferrer-i-Carbonell, and Muskens (2004) analyzed the changes in electricity and gas demands before and after the introduction of an environmental tax considering the ownership of several electronic appliances and found that home electronics affected energy demand. Rehdanz (2007) used panel data from 12,000 German households and analyzed the factors in space-heating expenditure. She found that energy-efficient heating systems and insulated windows significantly decreased the use of heating energy (i.e. gas and oil). She also found that owner-occupiers tended to invest in energy-saving equipment more than renters did and that owner-occupiers were more insensitive to changes in energy price than renters were. Finally, she found that owner-occupiers tended to prefer energy-efficient investment over energy-saving behavior.

To sum up, previous studies on energy-efficient investment and energy-saving behavior generally found that households that invest in energy-efficient equipment are high income and owner occupied and tended to not engage in energy-saving behavior. The main contribution of this study to the existing literature is its focus on the interaction of two policies: the EPP and the power-saving request. While many studies have investigated the roles of energy-saving investment and energy-saving behavior, less attention has been paid to the role of policy. By using 327 responses to a questionnaire survey conducted in the summer of 2012, this study analyzes actual energy savings by Japanese households, focusing on the role of their air conditioner purchase under the EPP. While several studies have examined the effect of the Japanese government's power-saving request during 2011 and 2012 (Tanaka and Ida 2013; Nishio and Ofuji 2014), the households' consumption data in these studies were self-reported and the effect of the power-saving request was not thoroughly examined in their regression model. Our study addresses these limitations by using the electricity bill to help ensure the objectivity of the power consumption data and by including a dummy variable to capture the influence of the power-saving request in two regions, namely Kansai and Shikoku.

6.4. Empirical analysis

6.4.1. Data

We administered a questionnaire survey to households in the Osaka prefecture and Matsuyama city in the summer of 2012. We asked the Nippon Research Center and the Matsuyama Chamber of Commerce and Industry to implement the survey among Osaka and Matsuyama households, respectively. The Nippon Research Center e-mailed a link to the survey on its web page to their monitors in the Osaka prefecture, while the Matsuyama Chamber of Commerce and Industry mailed a paper survey to the Matsuyama households (the Nippon Research Center distributed 400 questionnaires, and the Matsuyama Chamber of Commerce and Industry distributed 300 questionnaires). The questionnaire asked about the households' socioeconomic characteristics, nature and extent of electricity-saving behavior, number and usage of air conditioners, and presence or absence of energy-efficient air conditioners (i.e. air conditioners recommended by the EPP). Out of the 700 questionnaires that were distributed, 360 were completed, representing a recovery rate of about 51.4 percent (34.5 percent for the Nippon Research Center and 74.3 percent for the Matsuyama Chamber of Commerce and Industry).

In addition to completing the questionnaire, we asked the households to send their electricity bill for August 2012. Every month, electricity bills are dropped in households' mailboxes by meter readers of the electricity company in each region.[7] The bill includes the amount of a household's electricity usage and fees for the present month and for the same month a year ago, which indicates the energy savings from the previous year. This information enabled us to analyze the households' behavior objectively. Table 6.4 presents the general statistics of the data for our study. The average electricity usage for August 2011 and 2012 (*ELEC_2011* and *ELEC_2012*) showed no significant difference, albeit the level of savings increased marginally from 2011 to 2012 (by about 2.2 percent).

Table 6.4 shows that energy-efficient air conditioners accounted for about 25 percent of the total number of purchases of appliances that were recommended by the EPP from May 2009 to March 2011. It also shows that a significant number of energy-efficient air conditioners that were not recommended by the EPP (i.e. air conditioners that had lower energy-saving performance than those recommended by the program) were purchased. Because these air conditioners were cheaper than the recommended energy-efficient air conditioners, many households still purchased them. We obtained this information by asking households through the survey whether the air conditioner they purchased was recommended by the EPP. From this information, we found that the air conditioners that were not recommended by the EPP were also sufficiently energy efficient. With the revision of the Act on Temporary Measures for Promotion of Rational Uses of Energy and Recycled Resources in Business Activities in 1999, the Top Runner program[8] was introduced. Under this program, all air conditioner manufacturers must raise the energy efficiency of their air conditioners yearly, regardless of whether their air conditioners were recommended by the EPP. Therefore, almost every air conditioner available for sale in Japan today is very energy efficient, even if it is not recommended by the program.

In order to verify the representativeness of our survey data, we compared the income and age distributions of our participating households and those of

Table 6.4 Summary and descriptive statistics of the variables in the analysis

Variable	Definition	Mean	Std Dev	Min	Max	Obs
ELEC_2012	Electricity consumption for August 2012 (kWh)	442.364	243.093	82.000	1724.000	360
ELEC_2011	Electricity consumption for August 2011 (kWh)	442.139	251.231	76.000	1807.000	360
REDUCTION	Power-saving rate for August 2012 (%, year-on-year)	2.216	19.272	-66.500	100.530	360
ECOP	Number of recommended air conditioners purchased	0.253	0.620	0.000	4.000	360
NON_ECOP	Number of non-recommended air conditioners purchased	0.122	0.491	0.000	5.000	360
OSAKA	Households in the Osaka prefecture (yes = 1, no = 0)	0.381	0.486	0.000	1.000	360
AGE	Age of respondent	44.657	10.969	22.000	75.000	350
INCOME[1]	Income level	3.187	1.485	1.000	8.000	337
DUAL_INCOME	Dual-income household	0.592	0.492	0.000	1.000	360
DETACHED	Single-family house	0.534	0.500	0.000	1.000	356
OWN	Homeownership (yes = 1, no = 0)	0.674	0.469	0.000	1.000	356
FAM_SIZE[2]	Family size	2.898	1.302	1.000	7.000	354
CHILD[3]	Number of children under elementary school age	0.201	0.524	0.000	3.000	354
OLD[4]	Number of elderly people	0.240	0.560	2.000	0.000	354
NEP	Environmental awareness	16.092	3.083	7.000	25.000	359
ALT	Altruism	16.600	2.558	7.000	23.000	359
AIR_N	Number of air conditioners owned	3.567	1.501	1.000	9.000	358

AIR_TIME	Average daily usage time of main room air conditioner	6.269	5.059	0.000	24.000	342
AIR_WIDE	Size of room where main air conditioner is installed (square meters)	19.361	8.162	1.824	45.601	346
AIR_TEMP	Preset temperature of main room air conditioner (°C)	27.443	1.426	20.000	30.000	346
TEMP_2012	Average outside air temperature for August 2012 (°C)	28.505	0.5225417	29.473	27.68	357
TEMP_2011	Average outside air temperature for August 2011 (°C)	27.840	0.6881589	29.038	26.494	357

1 Under 2 million yen = 1, 2–4 million yen = 2, 4–6 million yen = 3, 6–8 million yen = 4, 8–10 million yen = 5, 10–12 million yen = 6, 12–14 million yen = 7, over 14 million yen = 8.

2 One = 1, two = 2, three = 3, four = 4, five = 5, six = 6, seven = 7, eight = 8, nine = 9, over ten = 10.

3 Zero = 0, one = 1, two = 2, three = 3, four = 4, over five = 5.

4 Zero = 0, one = 1, two = 2, three = 3, four = 4, over five = 5.

households in Japan as a whole. Figure 6.3 shows that both the income distributions of the sample households and of households in the entire Japan are skewed to the right. The shares of households in lower income categories are smaller in our sample than that of households in the entire Japan, while the shares of households in higher income categories are larger in our participating households than that in households in the entire Japan. These results suggest that the income level of the households in our study is slightly higher than that of

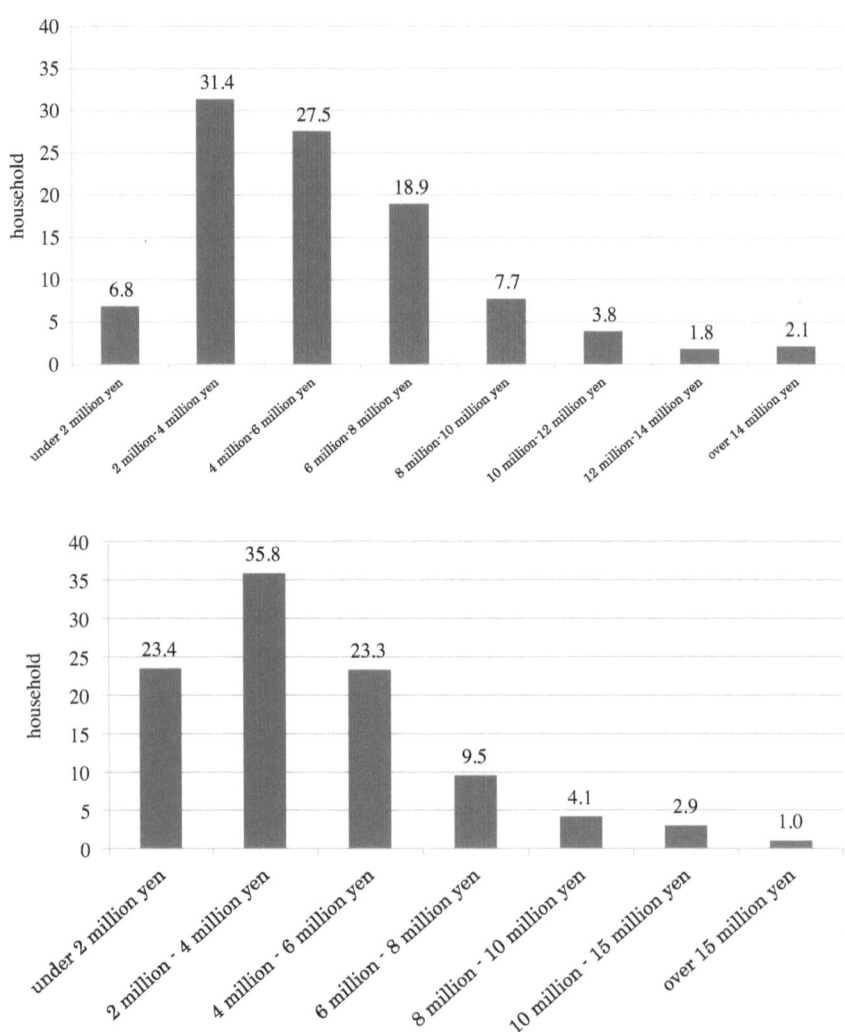

Figure 6.3 Income distributions (study sample)

Data source: National Tax Agency (2014).

households in the entire Japan. Figure 6.4 shows a significant difference between the age distribution of our participating households and that of households in the entire Japan. While most age groups in Japan accounted for about 8 percent of the entire population, the age distribution of our participating households followed a bell-shaped curve, peaking with the 40–44 years age group (19.1 percent). Moreover, the population shares of the 20–24 years and over 70 years groups were less than 2 percent. This means that the sample households were composed mainly of middle-aged working households and not so much of elderly and young households. This difference in the age distributions may have

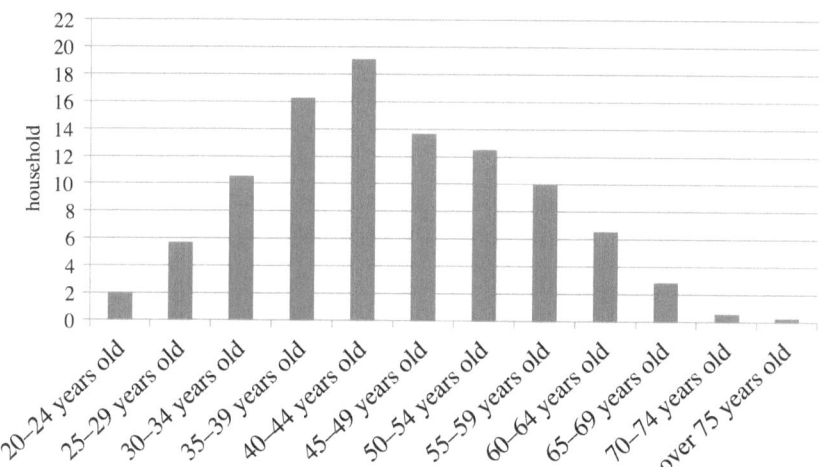

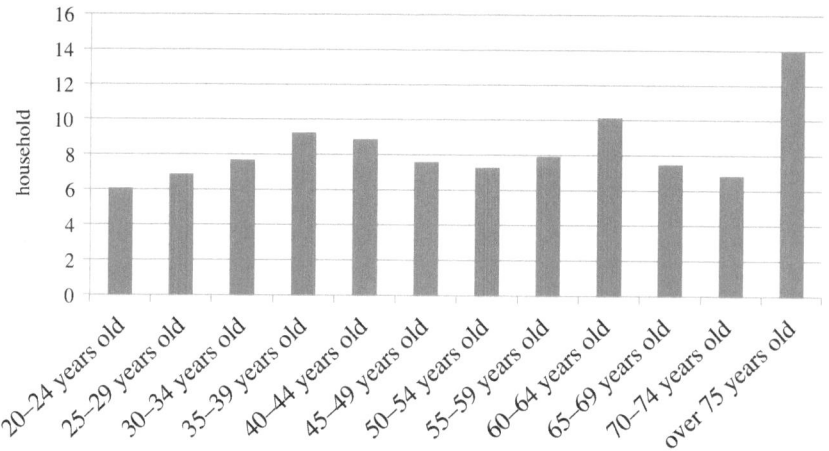

Figure 6.4 Age distributions (study sample)
Data source: Statistics Bureau of Japan (2014).

come from the characteristics of the research agency and the investigation method employed in our study. The households surveyed by the Matsuyama Chamber of Commerce and Industry were those of workers in its member companies, which thus excluded households of the unemployed and retired. Moreover, the Web-based questionnaire implemented by the Nippon Research Center may have inadvertently excluded elderly households who do not use the Internet. In summary, our participating households had a higher income level than households in the entire Japan (even though both groups of households have similar income distributions) and were composed mainly of middle-aged households rather than young and elderly households, compared to the households in the entire Japan.

Households from the Osaka region comprised about 40 percent of our sample. The Japanese government had requested these households to decrease their electricity consumption by 10 percent from the previous year[9] and had implemented a number of measures to help the households reach this goal. For example, local governments created a booklet and a web page about power saving, and the media were asked to encourage power saving. Electricity companies also implemented power-saving measures. For instance, the Kansai Electric Power Company implemented an electricity-saving trial, in which households that saved a certain amount of electricity were given a monetary coupon. In contrast, in Matsuyama households were requested to decrease their power usage only by 5 percent, and fewer power-saving measures were implemented.

In addition to their income and age, households provided their socioeconomic characteristics in the survey, such as whether they were a dual-income household, whether their house was a single-family house, the type of home ownership, family size, number of children under elementary school age, and number of elderly people. Moreover, the survey included questions about the households' environmental awareness and altruism, which potentially affected the households' power-saving behavior. We used the New Ecological Paradigm (NEP) scale, an index of environmental attitude that is commonly used in the social and behavioral sciences (Dunlap et al. 2000). Recently, the NEP scale has also been used in the economic literature on green household electricity programs (Kotchen and Moore 2007). The households were asked the extent to which they agreed or disagreed with the statements in the questionnaire through a 5-point Likert scale. Five statements from the NEP scale were included in our survey and combined into a summated scale that provides a measure of general environmental attitude (see Table 6.5). To measure household altruism, the respondents were asked to indicate the extent to which they agreed or disagreed with five statements from Schwartz (1970; 1977) on a 5-point Likert scale. Then, the five statements were combined into a summated scale that provides a measure of general altruistic attitude. Although this scale has been commonly used in laboratory experiments to explain the private provision of public goods, it has been used less frequently in field studies (Kotchen and Moore 2007). Our study used five statements from the altruism scale that provides a measure of altruistic attitude (see Table 6.5).

Table 6.5 Statements in the NEP and altruistic scales and their response distributions (percentage)

	Strongly agree	Partly agree	Unsure	Partly disagree	Strongly disagree	Correlation
NEP scale						
1. Human ingenuity will ensure that we do not make the earth unlivable.	94	141	73	25	14	0.372
2. Plants and animals have as much right as humans to exist.	94	67	67	113	7	0.432
3. The balance of nature is strong enough to cope with the impacts of modern industrial nations.	44	88	94	82	40	0.682
4. The so-called "ecological crisis" facing humankind has been greatly exaggerated.	16	50	104	109	69	0.681
5. The earth is like a spaceship with very limited room and resources.	116	142	72	16	2	0.480
Cronbach's coefficient alpha						0.358
Altruism scale						
1. Contributions to community organizations rarely improve the lives of others.	37	91	171	40	9	0.510
2. The individual alone is responsible for his or her well-being in life.	127	154	59	6	2	0.361
3. It is my duty to help other people when they are unable to help themselves.	30	52	195	63	8	0.572
4. My responsibility is to provide only for my family and myself.	12	56	142	79	59	0.585
5. My personal actions can greatly improve the well-being of people I don't know	43	151	122	25	7	0.672
Cronbach's coefficient alpha						0.398

Finally, we also asked the households about their ownership and usage of air conditioners. The households owned 3.6 air conditioners on average, higher than the average of 2.64 air conditioners owned by Japanese households (Energy Data and Modelling Center 2014). On average, the sample households installed main air conditioners in a room with an area of about 19.4 square meters and used them about 6.3 hours per day, with a preset temperature of 27.4°C.

6.4.2. Regression model

We estimated the household electricity demand function, taking electricity usage in August 2012 as a dependent variable, and examined the influence of energy-efficient investment. We considered whether the households purchased an energy-efficient air conditioner and examined its influence on the households' level of electricity saving in the summer. The regression model of our study is as follows:

$$\Upsilon_{2012,i} = \alpha + \beta_1 ECOP_i + \beta_2 NON\_ECOP_i + \gamma_1 OSAKA_i + \beta_3 X_i + \varepsilon_i$$

where $\Upsilon_{2012,}$ i indicates household i's electricity consumption in August 2012 (*ELEC_2012*). The variables $ECOP_i$ and NON_ECOP_i indicate the number of energy-efficient air conditioners owned that were recommended by the EPP and that were not recommended by the system, respectively. If an energy-efficient air conditioner is purchased to replace a less efficient unit, electricity consumption will decrease, provided that the household's consumption behavior does not change. Therefore, estimated coefficients of these variables are expected to be negative. However, as found in previous studies, households that invest in energy-efficient equipment engage in less energy-saving behavior than households that do not invest in such equipment (Rehdanz 2007; Meier and Rehdanz, 2010). Moreover, the government's power-saving request might have also promoted greater-than-usual energy saving among households that did not purchase such equipment. In short, the coefficients of $ECOP_i$ and NON_ECOP_i could be either positive or insignificant depending on these effects. $OSAKA_i$ is a dummy variable for whether a household is in the Osaka prefecture or not. This study targeted two regions, namely Kansai and Shikoku. Households in the Osaka prefecture belong to the Kansai region, while households in the Matsuyama prefecture belong to the Shikoku region. Table 6.3 shows that in the summer of 2012 households in Kansai were asked to reduce their power consumption by 10 percent, while those in Shinkoku were asked to reduce their power consumption by 5 percent. We expected that the higher level of electricity saving requested encouraged power-saving behavior, so that the estimated coefficient of $OSAKA_i$ may be negative. Finally, X_i represents the other socioeconomic variables shown in Table 6.4 (*AGE, INCOME, DUAL_INCOME, DETACHED, OWN, FAM_SIZE, CHILD, OLD, NEP, ALT, AIR_N, AIR_TEMP*, and *TEMP_2012*). We expected that the electricity demand in the summer depended largely on the usage of air conditioners, and therefore the estimated coefficients of the number of air conditioners (*AIR_N*) and the area of the room where the main air conditioner is installed (*AIR_WIDE*) may be positive.

6.4.3. Estimation results

We tested the heteroscedasticity of the error term with the Breusch-Pagan-Godfrey and White tests. The null hypotheses were rejected at the 1 percent significance level in both tests. Thus, we used the White heteroscedasticity-consistent standard errors to assess the heteroscedasticity assumption. Table 6.6 shows the estimation results. The first column indicates the results with the electricity consumption in 2012 as the dependent variable, whereas the second, third, and fourth columns show the results with the electricity consumption in 2011 and electricity-saving rate (2012 over 2011) as the dependent variable, respectively. The number of air conditioners (*AIR_N*) and the area of a room (*AIR_WIDE*) were positive and statistically significant. Thus, household electricity consumption in the summer was significantly affected by air conditioner use.

6.4.4. Effect of the power-saving request

We examined the effect of the electricity-saving request that was implemented in 2011 and 2012. As shown in Table 6.3, the electricity-saving request in the Kansai area (10 percent in both 2011 and 2012) was greater than in the Shikoku area (0 percent in 2011 and 5 percent in 2012). Therefore, household electricity-saving behavior was more strongly encouraged in Kansai than in Shikoku. All the estimated coefficients of the dummy for Kansai (i.e. Osaka) were negative and statistically significant. This indicates that households in the Osaka prefecture saved more electricity owing to the government's request. The Fukushima accident and the subsequent shutdown of other nuclear power plants in Japan have promoted power-saving awareness among Japanese residents, and these results indicate that the electricity-saving request could have further encouraged households to save electricity.

6.4.5. Effect of energy-efficient air conditioners

Next, we examined the effect of ownership of energy-efficient air conditioners on power saving. Table 6.6 shows that the estimated coefficients of *ECOP* and *NON_ECOP* were negative and statistically significant for the model with the electricity consumption for 2012 as the dependent variable. This result suggests that households that purchased energy-efficient air conditioners saved more electricity than households that did not. This result might also demonstrate that the EPP, which was aimed at reducing carbon dioxide emissions, played a role in reducing the energy deficit in Japan after the Fukushima crisis. In contrast, the estimation results in the second column of the same table, for the model with the electricity consumption for 2011 as the dependent variable, show that

Table 6.6 Estimation results

Variable	(1) ELEC_2012	(2) ELEC_2011	(3) REDUCTION	(4) REDUCTION (FGLS)
ECOP	−28.37 *	−17.85	−3.079 *	−3.440 *
	(16.85)	(16.39)	(1.620)	(1.894)
NON_ECOP	−37.39 **	−36.16 **	−1.491	−2.598
	(14.97)	(18.65)	(1.701)	(1.605)
OSAKA	−81.45 ***	−66.82 **	−9.099 ***	−8.816 ***
	(24.23)	(30.20)	(2.921)	(2.654)
AGE	0.546	1.132	−0.103	−0.164
	(1.136)	(1.262)	(0.107)	(0.104)
INCOME	11.03	9.644	1.347	0.931
	(9.452)	(11.43)	(0.886)	(0.840)
OWN	19.20	39.88	−4.292	−5.548 *
	(25.54)	(31.59)	(3.038)	(3.260)
DUAL_INCOME	−43.55 *	−60.4 **	3.728	4.237 *
	(22.39)	(26.60)	(2.631)	(2.408)
DETACHED	−70.79 ***	−76.33 **	−3.174	−2.668
	(26.39)	(32.17)	(2.881)	(2.944)
FAM_SIZE	56.71 ***	57.13 ***	−0.363	−0.006
	(11.16)	(13.40)	(1.064)	(1.056)
CHILD	−71.81 ***	−64.08 ***	−2.129	−2.414
	(19.87)	(20.84)	(2.454)	(2.219)
OLD	67.96 ***	60.2	1.020	0.529
	(21.96)	(21.11)	(1.723)	(1.679)
NEP	1.648	2.482	−0.407	−0.423
	(3.043)	(3.302)	(0.348)	(0.343)
ALT	−8.106 **	−5.862	−0.812 *	−0.706
	(4.040)	(4.320)	(0.485)	(0.443)
AIR_N	75.260 ***	73.000 ***	1.107	1.203
	(10.54)	(10.72)	(0.841)	(0.845)
AIR_WIDE	7.556 ***	7.351 ***	−0.0457	0.036
	(2.655)	(2.740)	(0.231)	(0.240)
TEMP_2012	12.84		7.189 ***	8.095 ***
	(19.95)		(2.248)	(2.076)
TEMP_2011		−5.268		
		(16.62)		
Constant	−289.5	152.6	−178.1 ***	−202.977 ***
	(580.7)	(480.7)	(64.34)	(59.98)
Observations	327	327	327	327
R-squared	0.496	0.452	0.114	0.0904

Standard errors are indicated in parentheses.

*** $p < 0.01$, ** $p < 0.05$, * $p < 0.1$.

the estimated coefficient of *NON_ECOP* was negative and statistically significant but that of *ECOP* was negative and statistically insignificant. Moreover, the estimated coefficient of *ECOP* in the first column was smaller than that of *NON_ECOP*, albeit the difference was not significant (we conducted the Wald test under the null hypothesis $\beta 1 = \beta 2$). To investigate the reasons for these results, we examined the usage of air conditioners among participating households during the study period.

Table 6.7 shows the estimation results of the regression analysis that take the average daily use time and the preset temperature of the main room air conditioner as the dependent variables. Because the demand for power in the summer depends largely on the usage of air conditioners, these two variables may be suitable indexes for the households' power-saving behavior in the summer. The table shows that for the power-saving behavior of households who purchased air conditioners (i.e. *ECOP* and *NON_ECOP*), all estimated coefficients were not significant, except for the estimated coefficient of the air conditioner usage time of households who purchased units recommended by the EPP. This result indicates that there was no large difference in power-saving behavior between households that purchased recommended air conditioners and those that did not. However, for the households that purchased recommended air conditioners, the coefficient of air conditioner usage time was positive and statistically significant. This shows that even though the power-saving request was implemented, these households did not perform any power-saving behavior, unlike other households. This is in line with the results of Rehdanz (2007) and Meier and Rehdanz (2010) that the price elasticity of energy demand of households that made energy-efficient investments was lower than those that did not. This finding suggests that households that made energy-efficient investments did not prefer energy-efficient behavior, unlike other households. To sum up, the EPP might have helped households save energy not by changing the purchasers' behavior but rather by the technological improvement of the equipment they purchased.

The rebound effect is another possible explanation for the increased air conditioner usage time among households that purchased units recommended by the EPP. Technological progress can reduce energy usage, which reduces the real cost of energy services per unit, which, in turn, increases the demand for energy services. Therefore, an energy reduction caused by a technological improvement might be partially offset by the cost reduction. Previous studies have referred to this phenomenon as the rebound effect (Sorrell and Dimitropoulos 2008). For example, let us consider a household that purchases a new air conditioner that consumes less energy than an older unit. If the new air conditioner's usage time and preset temperature are identical to those of the older model, energy consumption is decreased by this technological improvement. However, this improvement also reduces the operating costs, which increases the demand for air conditioning. Using the demand system model, Mizobuchi (2008) found a substantial rebound effect among Japanese households.

Table 6.7 Estimation results (effects on the average daily use time and the preset temperature of the main room air conditioner)

Variables	AIR_TIME	AIR_TEMP
ECOP	0.732 *	0.00270
	(0.441)	(0.125)
NON_ECOP	0.481	0.117
	(0.608)	(0.171)
OSAKA	0.339	1.073 ***
	(0.690)	(0.197)
AGE	−0.00572	0.00172
	(0.0280)	(0.00789)
INCOME	0.233	−0.0127
	(0.201)	(0.0570)
OWN	−2.224 ***	0.0153
	(0.789)	(0.221)
DUAL_INCOME	−1.908 ***	−0.244
	(0.644)	(0.182)
DETACHED	−0.00275	0.0399
	(0.741)	(0.209)
FAM_SIZE	0.529 *	−0.0732
	(0.277)	(0.0788)
CHILD	−0.290	0.329 *
	(0.629)	(0.178)
OLD	−0.0488	0.0579
	(0.528)	(0.149)
NEP	0.0935	0.0112
	(0.0935)	(0.0264)
ALT	−0.277 **	0.0607 ***
	(0.118)	(0.0334)
AIE_WIDE	0.208 **	0.0150
	(0.0669)	(0.0188)
TEMP_2012	0.427	−0.0433
	(0.563)	(0.160)
Constant	−4.650	27.08 **
	(16.17)	(4.588)
Observations	320	316
R-squared	0.140	0.156

Standard errors are indicated in parentheses.

*** $p < 0.01$, ** $p < 0.05$, * $p < 0.1$.

6.4.6. *Effect of other socioeconomic variables*

Regarding the other socioeconomic variables in Table 6.6 and Table 6.7, the factors that had a positive and significant effect on electricity consumption were family size (*FAM_SIZE*) and number of elderly people (*OLD*). As shown in Table 6.7, the positive and significant estimated coefficient of *FAM_SIZE*, whose dependent variable is usage time (*AIR_TIME*), suggests that the greater the number of people in a household in the summer, the higher the electricity demand. In contrast, factors that had a negative and significant effect on electricity consumption were being a dual-income household (*DUAL_INCOME*), having a single-family house (*DETACHED*), the number of children under elementary school age (*CHILD*), and altruism (*ALT*). In a dual-income household, household members spend less time at home on weekdays, so their air conditioner usage time may be less than that of other households. The negative and significant estimated coefficient of the dual-income household variable in the first column of Table 6.7 supports this expectation, too. The positive and significant estimated coefficient of the number of elementary school children (*CHILD*) in the second column of Table 6.7 also supports the expectation that parents increase the air conditioner's preset temperature for the sake of the children's health or metabolism. In Table 6.7 only altruism (*ALT*) had a significant effect on both *AIR_TIME* and *AIR_TEMP*, which shows that altruistic households saved more electricity (decreased air conditioner usage time and increased preset temperature) than non-altruistic households. Without the households' compliance with the government's power-saving request, large-scale power outages would have occurred in the summer of 2012. This scenario prompted many households to save more electricity not only for their own benefit, but also for the benefit of others (i.e. they were altruistic households). Finally, we evaluated the effects of environmental awareness. We expected that households that were more environmentally aware would save more power, but the insignificant estimated coefficient for environmental awareness in Tables 6.6 and 6.7 suggests that environmental awareness did not promote energy-saving behavior in our sample. This result may be because households with higher environmental awareness already had lower electricity usage than other households to begin with, which suggests that after the Fukushima accident they did not have to reduce their electricity consumption as much as other households.

6.5. Conclusions and suggestions for future studies

This study estimated Japanese households' electricity demand functions based on both electricity demand data for the summer of 2012 and questionnaire data. We analyzed the power-saving effects of energy-efficient air conditioners that were promoted by the EPP. Moreover, we divided these households into those that purchased energy-efficient air conditioners recommended by the program and those that purchased other air conditioners. The estimation results indicate that both groups of households significantly decreased power consumption. However, when we used the power consumption for the summer of 2011 as the dependent variable, we did not find a significant saving effect of the purchase of recommended air conditioners. This may be because, as shown in previous studies, households that made energy-efficient investments may not perform energy-saving behavior, in contrast to households that did not make energy-efficient investments. That is, there is a trade-off between energy-efficient investment and energy-saving behavior. Moreover, as shown in Table 6.7, households that purchased recommended air conditioners had significantly longer daily usage time than other households. This suggests a rebound effect of the purchase of energy-efficient equipment. In addition, our results demonstrate that the level of power saving of households during the implementation of the government's electricity-saving request depended on whether they purchased recommended air conditioners or purchased other air conditioners and on the air conditioners' energy efficiency level. From these findings, we conclude that additional power-saving measures must be considered to further encourage power saving among households that purchased recommended energy-efficient air conditioners.

To analyze the energy-saving behavior of households that purchased energy-efficient air conditioners, it is important to consider the reason for their purchase – to replace an old air conditioner with a new, efficient air conditioner or to purchase an additional air conditioner. Generally, if a household replaces an old air conditioner with an energy-efficient one and maintains the same level of usage, it reduces its energy consumption because of the higher performance of the new air conditioner. On the other hand, if a household purchases a new air conditioner in addition to an existing one, it increases its level of usage and thus increases its total power consumption. To the best of our knowledge, no study has examined the energy-saving effect of energy-efficient home appliances considering households' reason for purchasing. In Japan, many households use room air conditioners to control temperature, so it is relatively easy to replace them or to purchase additional units. Future studies could examine power-saving behavior considering the reason for purchase in other geographical contexts.

Requesting a higher level of electricity saving is a potential measure to further encourage power-saving behavior. In our study, significant power-saving effects were observed in the Osaka prefecture, where a higher level of electricity saving was requested. Further, the cost-efficiency of an electricity-saving request should be compared with other policy measures, such as varying the electricity price by time zone (Tanaka and Ida 2013), providing monetary rewards (Mizobuchi and Takeuchi 2012; 2013), and appealing to altruistic motivations.

Acknowledgments

Author Kenichi Mizobuchi thanks the Japan Society for the Promotion of Science (Grant-in-Aid for Young Scientists [B] #22730217) for supporting this research. Author Kenji Takeuchi thanks the Department of Economics at the University of Gothenburg for giving him the opportunity to be a visiting researcher.

Notes

1 The recommended products included three types of electronic appliances – air conditioners, refrigerators, and terrestrial digital TVs – that are very energy efficient.
2 The total number of eco points awarded in May 2011 was about 44.81 million (METI et al. 2010).
3 All 54 nuclear power reactors in Japan were shut down by May 2012 and remained closed as of December 2014.
4 Following the Agency for Natural Resources and Energy's 2009 Civilian Sector Energy Consumption Survey, to the question "Is the appliance you purchased within the year bigger than old one?" 84.7 percent of terrestrial digital TV purchasers and 69 percent of refrigerator purchasers replied "Yes." Additionally, about 41 percent of respondents who bought a new terrestrial digital TV during the implementation of the EPP mentioned that their power consumption increased compared with the previous year. These results suggest that many consumers might have been encouraged to purchase more home electronics that consumed a significant amount of power because of the significant number of eco points that came with them (Board of Audit of Japan 2012).
5 The 10 largest power companies of Japan are Hokkaido Electric Power Co., Inc.; Tohoku Electric Power Co., Inc.; Tokyo Electric Power Co., Inc.; Hokuriku Electric Industry Co., Ltd.; Chubu Electric Power Co., Inc.; Kansai Electric Power Co., Inc.; Chugoku Electric Power Co., Inc.; Shikoku Electric Power Co., Inc.; Kyushu Electric Power Co., Inc.; and Okinawa Electric Power Co., Inc. The bill for the full liberalization of the electricity retail market was passed on 11 June 2014 and will be implemented around 2016. The structure of the power supply of Japan is expected to vary greatly after the implementation of the bill (see Revision of the Electricity Utilities Industry Law).
6 Because there are no nuclear plants in Okinawa, the power-saving request was not implemented there. Moreover, in view of the power interchange between areas, the level of power-saving requested of each area was somewhat high.
7 There are 10 large electricity companies that are exclusive suppliers for each region in Japan. The Osaka prefecture is supplied by the Kansai Electric Power Co. (http://www1.kepco.co.jp/english/), and Matsuyama city is supplied by the Shikoku Electric Power Co. (http://www.yonden.co.jp/english/index.html).
8 This program was created to promote the manufacture of energy-efficient electronic appliances and automotive vehicles. This program set energy-efficient targets based on the value of the most energy-efficient products in the market, which all machinery and equipment covered by the program should exceed. Companies that continue to manufacture and sell products that fail to meet the targets are publicized and penalized (http://www.enecho.meti.go.jp/policy/saveenergy/toprunner2011.03en-1103.pdf).
9 Until the Ohi nuclear power plants 3 and 4 of the Kansai Electric Power Co. resumed operations on July 2012, households were asked to reduce electricity consumption by 15 percent. However, after operations resumed, this request was reduced to 10 percent.

References

Agency for Natural Resources and Energy. (2009) 'Energy Consumption Survey in 2009 (in Japanese)', Tokyo.

H. Arikawa, Y. Cao, and S. Matsumoto. (2014) 'Attitudes about nuclear power and energy-saving behavior among Japanese households', *Energy Research & Social Science* 2, 12–20.

S. Banfi, M. Farsi, M. Filippini, and M. Jakob. (2008) 'Willingness to pay for energy-saving measures in residential buildings', *Energy Economics* 30, 503–16.

P.H.G. Berkhout, A. Ferrer-i-Carbonell, and J.C. Muskens. (2004) 'The ex post impact of an energy tax on household energy demand', *Energy Economics* 26, 297–317.

Board of Audit of Japan. (2012) 'The effect of subsidies on the diffusion of green home electronics (in Japanese)', Tokyo. Available: <http://www.jbaudit.go.jp/pr/kensa/result/24/pdf/241011_zenbun_1.pdf> (accessed 16 July 2014).

E.R. Branch. (1993) 'Short run income elasticity of demand for residential electricity using consumer expenditure survey data', *Energy Journal* 14, 111–21.

T.A. Cameron. (1985) 'A nested logit model of energy conservation activity by owners of existing single family dwellings', *Review of Economics and Statistics* 67, 205–11.

D. Charlier. (2013) 'Energy-saving investments in the residential sector: an econometric analysis', paper presented at European Association of Environmental and Resource Economists 20th Annual Conference, Toulouse, June.

J.A. Dubin and D.L. McFadden. (1984). 'An econometric analysis of residential electric appliance holdings and consumption', *Econometrica* 52, 345–62.

R.E. Dunlap, K.D. Van Liere, A.D. Mertig, and R.E. Jones. (2000). 'Measuring endorsement of the new ecological paradigm: a revised NEP scale', *Journal of Social Issues* 56, 425–42.

Energy Data and Modelling Center. (2014) *Handbook of Energy & Economic Statistics* (in Japanese), Tokyo: The Institute of Energy Economics, Japan, Quantitative Analysis Unit.

C. Garbacz. (1985) 'Residential fuel oil demand: a micro-based national model', *Applied Economics* 17, 669–74.

R.D. Green. (1987) 'Regional variations in US consumer response to price changes in home heating fuels: the Northeast and the South', *Applied Economics* 19, 1261–8.

Greenhouse Gas Inventory Office of Japan. (2014) 'The GHGs Emissions Data of Japan', Tokyo. Available: <http://www-gio.nies.go.jp/aboutghg/nir/nir-e.html> (accessed 4 July 2014).

P. Grosche and C. Vance. (2009) 'Willingness-to-pay for energy conservation and free-ridership on subsidization: evidence from Germany', *Energy Journal* 30, 135–54.

B. Halvorsen and B.M. Larsen. (2001) 'The flexibility of household electricity demand over time', *Resource and Energy Economics* 23, 1–18.

T.Y. Jung. (1993) 'Ordered logit model for residential electricity demand in Korea', *Energy Economics* 15, 205–9.

M.J. Kotchen and M.R. Moore. (2007) 'Private provision of environmental public goods: household participation in green-electricity programs', *Journal of Environmental Economics and Management* 53, 1–16.

R.J. Krumm. (1983) 'Durable good choice: a benefit-cost approach to air conditioning', *Resources and Energy* 5, 369–401.

S. Leth-Petersen and M. Togeby. (2001) 'Demand for space heating in apartment blocks: measuring effects of policy measures aiming at reducing energy consumption', *Energy Economics* 23, 387–403.

K. Mahapatra and L. Gustavsson. (2008) 'An adopter-centric approach to analyze the diffusion patterns of innovative residential heating systems in Sweden', *Energy Policy* 36, 577–90.

H. Meier and K. Rehdanz. (2010) 'Determinants of residential space heating expenditures in Great Britain', *Energy Economics* 32, 949–59.

R. Mendelsohn. (1977) 'Empirical evidence on home improvements', *Journal of Urban Economics* 4, 459–68.

Ministry of Economy, Trade and Industry, Ministry of the Environment and Ministry of Internal Affairs and Communications. (2010) 'Policy effects of the consumer electronics Eco-point system (in Japanese),' Tokyo. Available: <http://www.meti.go.jp/press/2011/06/20110614002/20110614002-2.pdf> (accessed 16 July 2014).

K. Mizobuchi. (2008) 'An empirical study on the rebound effect considering capital costs', *Energy Economics* 30, 2486–516.

K. Mizobuchi and K. Takeuchi. (2012) 'Using economic incentive to conserve electricity consumption: a field experiment in Matsuyama', *International Journal of Energy Economics and Policy* 2, 318–32.

K. Mizobuchi and K. Takeuchi. (2013) 'The influences of financial and non-financial factors on energy-saving behaviour: a field experiment in Japan', *Energy Policy* 63, 775–87.

C. Montgomery. (1992) 'Explaining home improvement in the context of household investment in residential housing', *Journal of Urban Economics* 32, 326–50.

G. Nair, L. Gustavsson, and K. Mahapatra. (2010) 'Factors influencing energy efficiency investments in existing Swedish residential buildings', *Energy Policy* 38, 2956–63.

National Tax Agency. (2014) Available: <http://www.nta.go.jp/index.htm> (accessed 4 July 2014).

R. Nesbakken. (1999) 'Price sensitivity of residential energy consumption in Norway', *Energy Economics* 21, 493–515.

R. Nesbakken. (2001) 'Energy consumption for space heating: a discrete-continuous approach', *Scandinavian Journal of Economics* 103, 165–84.

K. Nishio and K. Ofuji. (2013) 'Ex-post analysis of electricity saving measures in the residential sector in the summer of 2012 (in Japanese)', Tokyo: Central Research Institute of Electric Power Industry.

K. Nishio and K. Ofuji. (2014) 'Ex-post analysis of electricity saving measures in the residential sector in the summer of 2013 (in Japanese)', Tokyo: Central Research Institute of Electric Power Industry.

K. Rehdanz. (2007) 'Determinants of residential space heating expenditures in Germany', *Energy Economics* 29, 167–82.

S.H. Schwartz. (1970) 'Elicitation of moral obligation and self-sacrificing behaviour: an experimental study of volunteering to be a bone marrow donor', *Journal of Personality and Social Psychology* 15, 283–93.

S.H. Schwartz. (1977) 'Normative influences on altruism', in L. Berkowitz (ed.), *Advances in Experimental Social Psychology* 10, New York: Academic Press.

S. Scott. (1997) 'Household energy efficiency in Ireland: a replication study of ownership of energy saving items', *Energy Economics* 19, 187–208.

S. Sorrell and J. Dimitropoulos. (2008) 'The rebound effect: microeconomic definitions, limitations and extensions', *Ecological Economics* 65, 636–49.

Statistics Bureau of Japan. (2014) Available: <http://www.stat.go.jp/index.htm> (accessed 4 July 2014).

M. Tanaka and T. Ida. (2013) 'Voluntary electricity conservation of households after the Great East Japan Earthquake: a stated preference analysis', *Energy Economics* 39, 296–304.

K. Vaage. (2000) 'Heating technology and energy use: a discrete/continuous choice approach to Norwegian household energy demand', *Energy Economics* 22, 649–66.

X. Wu, J. Lampietti, and A.S. Meyer. (2004) 'Coping with the cold: space heating and the urban poor in developing countries', *Energy Economics* 26, 345–57.

7 Effect of the eco-point program on the implicit discount rate

A hedonic analysis of the eco-point program

Shigeru Matsumoto, Minoru Morita, and Tomohiro Tasaki

7.1. Introduction

Global residential energy use is expected to grow at an average rate of 1.1 percent per year from 2008 to 2035 (U.S. Energy Information Administration 2011). The growth rate in non-OECD countries will be much higher than that in OECD countries. Currently, most countries have policies and programs to promote energy-efficient products.

If market forces were fully effective and consumers chose their products rationally, then such policies and programs would not be necessary. Consumers would compare the upfront cost with the lifetime energy reduction and choose energy-efficient products. However, typical consumers underestimate the benefits of future energy saving and underinvest in energy efficiency relative to a description of the socially optimal level of energy efficiency. This phenomenon is called the "energy efficiency gap," and it has been studied for decades.[1]

One explanation for the energy efficiency gap is asymmetric information. Consumers often lack sufficient information about the difference in future operating costs between more efficient and less efficient products to make proper investment decisions (Howarth and Sanstad 1995). If they are properly informed about the operating cost savings, then they may invest in energy efficiency. Product labels are utilized in many countries to provide consumers with recognizable information on home electric appliances.[2]

Previous studies, however, have demonstrated that consumers do not choose energy-efficient products even if they are informed of the benefit of future energy saving. Consumers underestimate the benefit of energy-saving investments. The literature reports that implicit discount rates for energy-saving investments range from 25 percent to over 100 percent (Sanstad, Hanemann, and Auffhammer 2006).

To alter consumers' myopic behavior, many countries have implemented energy efficiency programs. For instance, the U.S. government uses minimum energy efficiency standards for major appliances, and manufacturers in the United States must produce products that meet the standard. Similarly, the

Japanese government uses the Top Runner program, which sets the energy efficiency standard based on the most efficient product on the market. Producers must ensure that the weighted average energy efficiency of the products they sell in the target year achieves the requisite standards (Kimura 2010).

By raising the energy efficiency standard, the government can drive inefficient products out of the market gradually. However, it cannot provide consumers with an incentive to select an energy-efficient product in the present market. If consumers are seriously concerned about the upfront costs, they choose energy-inefficient products.

To provide consumers with an incentive to select an energy-efficient product, many countries have started to implement rebate programs for energy-efficient products. In the presence of temporal myopia, such rebate programs could be more effective than energy taxes that raise operating costs over time. The purpose of this chapter is examine how energy-efficient rebate programs affect consumer valuation of energy efficiency.[3]

The eco-point program for home electric appliances in Japan provides an excellent opportunity for this purpose.[4] Most rebate programs decrease the tax for efficient products and increase it for inefficient ones. In contrast, in the eco-point program, consumers who purchased energy-efficient home electric appliances obtained eco points that could be used to buy other goods and services.[5] This unique rebate system provided a strong incentive for consumers to choose energy-efficient appliances.

The eco-point program (EPP) started on 15 May 2009 and ended on 31 March 2011. The Ministry of Economy, Trade and Industry (2011) estimated that the shipment of home electric appliances increased by 67 percent and the program expanded the domestic economy by 2.6 trillion yen ($21.67 billion) during this period.[6] Therefore, the impact of the program on the economy was significant (see further detail in Chapter 6).

The products covered by the system and the points provided to consumers were modified several times. Therefore, within a relatively short period of time acquisition prices of electric appliances were modified according to the energy efficiency criteria. Thus, we can ignore the effect of technology innovation and focus on consumer valuation of energy saving.

Finally, the residential electricity price in Japan is high, and a wide variety of energy-efficient appliances are available in the Japanese market.[7] The market is quite competitive. These market and policy conditions make the analysis of the EPP particularly interesting.

The appliances covered by the EPP were air conditioners, refrigerators, and televisions. In this chapter, we focus on air conditioners for the following two reasons. First, air conditioners sold in the Japanese market are used for both space cooling and heating, and thus are the home electric appliances that consume the largest amount of electricity.[8] Second, previous research on the impact of energy efficiency regulations has focused on air conditioners (Hausman 1979; Ruderman, Levine, and McMahon 1987). Therefore, we can compare our findings with those of previous studies.

   The rest of the chapter is organized as follows. In Section 7.2, we summarize the findings from previous research and then explain the Japanese eco-point program. In Section 7.3, we discuss the data and specify the empirical model. The purpose of this chapter is to describe how the rebate program affected consumer valuation of energy efficiency. We will conduct this task based on the hedonic price framework. Section 7.4 presents the empirical findings. The results show that consumers placed a lower value on energy saving during the eco-point period. Based on the empirical estimation, the EPP roughly doubled the implicit discount rate. Section 7.5 concludes the chapter.

7.2. Background

7.2.1. Related research

The energy efficiency gap is often illustrated by the implicit discount rate; that is, the rate consumers use to compare future appliance operating cost savings against an appliance purchase price premium. Implicit discount rates of various durables have been estimated over the last four decades. Although all literature shows that a typical consumer applies a high discount rate for energy investment, the estimated rate varies across durables. With regard to air conditioners, Hausman (1979) estimated the implicit discount rate at about 20 percent, while Ruderman, Levine, and McMahon (1987) estimated the rate to range from 19 to 22 percent.

   A variety of empirical models have been used for the estimation of implicit discount rates of home electric appliances. The major difficulty in the empirical estimation is that the price of electric appliances is not linear with respect to the energy efficiency measure. To overcome this nonlinearity problem, Hausman (1979) applied a qualitative choice model for the analysis of the survey data. Revelt and Train (1998) used a mixed logit model for the estimation of the implicit discount rate of refrigerators. Other papers have considered a nonlinear specification. For example, Greening, Sanstad, and McMahon (1997) include quadratic terms of the energy efficiency variable in the hedonic pricing model of refrigerators.

   The model selection in energy efficiency analysis is inconclusive at present. Some scholars apply choice models for the survey data, while others apply hedonic pricing models for the market data. In this study, we take the latter approach. To overcome the nonlinearity problem of the energy efficiency measure, we include room size in the empirical model.

7.2.2. Eco-point program for appliances

The government of Japan introduced the eco-point program with three objectives: (1) to reduce CO_2 emissions by encouraging consumers to choose energy-saving electric appliances, (2) to spread TVs compatible with digital broadcasting, and (3) to mitigate the negative impact of the Lehman shock by stimulating domestic consumption.

Table 7.1 Eco and recycle points provided for air conditioners

Cooling Capacity (Kilowatt-hour)	15 May 2009– 30 November 2010	1 December 2010– 31 March 2011
Above 3.6	9,000	5,000
2.8, 2.5	7,000	4,000
Below 2.2	6,000	3,000
Recycle Points	3,000	3,000

The EPP started on 15 May 2009 and ended on 31 March 2011 (See Figure 4.1 on page 78). During this period, the program was modified several times. The number of points provided was reduced on 30 November 2010. The energy efficiency criteria were tightened on 31 December 2010. In addition to the eco points, recycle points were provided for the replacement of old appliances until 31 December 2010.

The appliances covered by the EPP were air conditioners, refrigerators, and televisions with energy-saving labels. The points consumers obtained varied by type and size of appliance. Table 7.1 shows the eco and recycle points provided for air conditioners. As the table shows, more points were provided for larger air conditioners.

During the EPP period, the acquisition price (the net upfront cost) of air conditioners was modified several times according to the energy efficiency criteria. In the remaining sections, we will examine how the EPP affected consumer valuation of the energy efficiency measure of air conditioners.

7.3. Data and empirical model

7.3.1. Data

The primary data are sales data collected by the market research firm GfK Marketing Service Ltd. Japan. Additional information about the GfK Group is provided in Chapter 4. The data on air conditioners cover about 55 percent of total annual sales in Japan. The total value and number of sales of each product are reported by month. We calculate the average sales price of each product by dividing the total value of sales by the total number of sales.[9] We use this average sales price for the following hedonic analysis. In addition to price information, the detailed specifications of the products are available from the dataset. We include them as control variables.

Each air conditioner is designed to fit a specific room size.[10] It is important to choose an air conditioner with appropriate cooling capacity (room size), since an incorrectly sized air conditioner with insufficient or excess cooling capacity will result in inefficiency.

Air conditioners designed for larger rooms require more electricity and are more expensive in general. To measure consumer valuation of energy efficiency precisely, we have to control for the room size of air conditioners. Otherwise, we will simply conclude that energy-inefficient products are more expensive.

Table 7.2 provides summary statistics of air conditioners according to their designed room size. The first column shows the room size, and the second column shows the corresponding cooling capacity. As the third column of the table shows, the price of air conditioners increases as the room size increases. Therefore, there is a positive correlation between room size and price for air conditioners.

7.3.2. Empirical model

We employ the hedonic framework proposed by Rosen (1974) for the empirical analysis. An individual derives utility from a vector of air conditioner attributes, $X = x_1, . . ., x_n$. The price of the air conditioner is given by the function of attributes, $P = P(X)$. Marginal attribute price, $p(x_k)$, is defined as the partial derivative of price with respect to the attribute, $p(x_k) = \partial P/\partial x_k$.

Following Dreyfus and Viscusi (1995), we specify the equilibrium hedonic price locus as

$$P = \beta_0 + \beta_1 \left[Room\, Size_i \right]^\lambda + \beta_2 \left[Energy\, Efficiency_i \right] + \Sigma_k \beta_k A_{ki}^\lambda + \Sigma_l \beta_l A_{li} + \varepsilon_i$$

where λ is the Box-Cox transformation coefficient, which is interpreted as follows: $A_{ki}^\lambda = \left(A_{ki}^\lambda - 1 \right)/\lambda$ for $\lambda \neq 0$ or $A_{ki}^\lambda = \log A_{ki}$ for $\lambda = 0$. In this specification, the first bracket term measures consumer valuation of space cooling while the second term measures consumer valuation of energy efficiency.

We use annual performance factor (APF) as an energy efficiency measure. It is the most popular measure in the Japanese market. Various consumer reports suggest that consumers check the APF when shopping. Manufacturers estimate the annual electricity consumption of each air conditioner based on the running conditions specified by the Japanese Industrial Standard. APF is calculated by dividing the required cooling capacity by the estimated electricity consumption:

$$\mathrm{APF} = \frac{Required\, Annual\, Cooling\, Capacity\, (kWh)}{Estimated\, Annual\, Cooling\, Capacity\, (kWh)}.$$

Hence, APF indicates the energy efficiency level for a given room size, and a higher APF value implies that the air conditioner is more energy efficient. Table 7.2 shows that APF varies within models for the same room size.

Manufacturers release new models almost every year. The price of an existing model decreases as new models are released. To distinguish between old and new models, we include the period of time after the initial sales date of each air conditioner in A_{ki}. We also include the physical size of the air conditioners.

Table 7.2 Summary statistics for air conditioners according to designed room size[a]

Room Size (m²)	Cooling Capacity (kW)	Price (yen)[b]				Power (kWh)				APF			
		Average	Std. Dev.	Min	Max	Average	Std. Dev.	Min	Max	Average	Std. Dev.	Min	Max
All		117,063	51,151	28,478	246,420	1,407	663	314	3,557	5.4	0.8	3.0	7.0
9	2.2	75,637	32,935	28,478	147,943	800	133	377	1,442	5.8	0.8	3.0	7.0
12	2.5	90,345	33,400	43,644	157,452	883	107	535	1,721	5.8	0.7	5.0	7.0
15	2.8	102,298	36,540	48,114	177,339	990	157	314	2,002	5.9	0.7	5.0	7.0
18	3.6	129,599	37,239	66,681	183,188	1,392	161	1,110	1,731	5.4	0.6	4.0	7.0
21	4.0	126,747	38,748	58,085	200,275	1,565	218	440	2,435	5.3	0.7	4.0	7.0
23	4.5	79,606	7,986	75,371	83,841	1,982	186	1,701	2,286	4.5	0.5	4.0	5.0
24	5.0	153,200	39,017	82,394	221,408	2,112	266	1,670	3,024	4.9	0.8	3.0	6.0
27	5.6	165,240	41,187	122,474	213,510	2,321	251	1,935	2,806	4.9	0.4	4.0	6.0
30	6.3	198,541	23,485	149,329	230,333	2,677	202	2,295	3,154	4.8	0.5	4.0	6.0
33	7.1	216,968	21,072	176,063	244,877	3,191	205	2,700	3,557	4.6	0.5	4.0	5.0

a Models with fewer than 50 sales are excluded from the dataset.

b $1 is approximately 120 yen.

Table 7.3 Descriptive statistics for air conditioners: control variables (All sample: N = 2,075)

	Unit	Mean or Share	Standard Deviation
Room size	m²	17.87	6.55
APF	Real value	5.40	0.72
Physical size	cm³	57.60	12.07
Time after model's release (duration)	Year	3.90	3.05
Manufacturer dummy			
Corona	%	3.18	
Sharp	%	7.04	
Panasonic	%	17.54	
MHI	%	4.58	
Mitsubishi	%	13.98	
Sanyo	%	3.95	
Toshiba	%	10.55	
Hitachi	%	12.72	
Fujitsu General	%	5.98	
Other functions dummy			
Ion/bacteria elimination	%	11.57	
Ion emission/bacteria elimination	%	8.19	
Ion emission	%	18.41	
Ion	%	7.18	
Bacteria elimination	%	20.34	
Other clarification/bacteria elimination	%	8.82	
Air control	%	37.06	

We include the dummy variables of air conditioner functions such as ion emission and bacteria elimination. Finally, we include nine manufacturer dummies in A_{ii}. Table 7.3 presents the descriptive statistics.

7.4. Empirical results

We estimated the hedonic pricing model for each month. Table 7.4 shows the estimation results in the case of July. The EPP was introduced on 15 May 2009. Therefore, the first model (July 2008) shows the empirical result before the program, while both the second model (July 2009) and the third model (July 2010) show the results after the program.

Table 7.4 Results of hedonic price estimations (July)

Sampling Year	2008	2009	2010
(Number of Observations)	(302)	(305)	(360)
Room size	225.67*	616.62*	559.20*
APF	14,798.92*	16,859.37*	14,465.91*
Physical size	19.55*	57.83*	57.64*
Time after model's release	–6,191.67*	–10,180.87*	–7,273.00*
Other functions			
Ion/bacteria elimination	27,565.90*	10,035.85	1,993.43
Ion emission/bacteria elimination	15,921.24*	15,617.12*	–2802.00
Ion emission	10,200.62	19,412.32*	26,150.91*
Ion	16,262.74*	8,957.40	13,152.10
Bacteria elimination	–14,671.83	–14,287.51*	–13,815.85*
Other clarification/ bacteria elimination	0.00	14,206.63*	25,194.39*
Air control	17,720.23*	19,938.10*	10,477.87*
Manufacturer dummy			
Manufacturer A	–24,929.38*	–18,634.45*	–13,310.95*
Manufacturer B	–1,603.05	–18,833.16*	–23,440.92*
Manufacturer C	10,667.56*	1,455.83	–5,370.79
Manufacturer D	–29,400.40*	–36,967.83*	–29,726.74*
Manufacturer E	1.03	–3,928.46	–8,464.17*
Manufacturer F	–14,215.56	0.00	0.00
Manufacturer G	8,595.04	12,056.49*	7,713.34*
Manufacturer H	3,341.34	3819.33	1,800.21
Manufacturer I	–20,261.77*	–20,943.76*	–11,328.84*
Constant	–48,525.53	–59,405.46	–58,438.19
Λ	2.05*	1.70*	1.73*
Likelihood rate	–3,326.25	–3,340.70	–3,950.15
χ^2-value	723.30*	747.98*	907.30*

Note: * indicates statistically significantly different from zero at the 5% level.

As the table shows, the price of the air conditioner increases with the inclusion of additional functions. For instance, the inclusion of air control function increases the price of air conditioners by 10,477.87–19,938.10 yen ($87.32–$166.15).

The table further shows that the price of air conditioners decreases after the model's release. The rate of price decline is very rapid. Manufacturers need to cut the price of air conditioners by 6,191.67–10,180.87 yen ($51.60–$84.83) per year. There is considerable price variation across manufacturers. The empirical results reveal that the EPP altered the brand value.

The coefficient of room size is positive and significant. As expected, the result implies that air conditioners designed for a larger room are more expensive. Since $\lambda > 1.0$, the price of the air conditioner increases exponentially with room size. Based on the estimation result of July 2008, the average price of an air conditioner for a 9 m² room (a typical child's bedroom) is 20,327.77 yen ($169.40), while that for a 24 m² room (a typical living room) is 151,571.22 yen ($1,263.09).

There are three peak sales periods for air conditioners: March, July, and December. Table 7.5 compares the coefficients of room size and APF across these three months.[11] It shows that the coefficient of room size for July is smaller than that for December. Consumers purchasing an air conditioner in July may use it for space cooling, while those purchasing it in December may use it for space heating. The results suggest that the value of space cooling is lower than the value of space heating.

The coefficients of APF are positive and statistically significant at the 5 percent level in all models. The results imply that the energy-saving feature of air conditioners is reflected in their prices.

Using the coefficient of APF, we can calculate the implicit discount rate. Consider an air conditioner designed for a 21 m² room. The electricity price is about 22 yen/kWh ($0.18/kWh) in Japan. If an individual installed an air conditioner with an APF value larger by one unit, he or she could save an annual electricity cost of 5,280.92 yen ($44.00). Suppose the coefficient of APF from Table 7.5 is 20,000 and the life span of air conditioners is 15 years; in this case, the implicit discount rate becomes about 26 percent.

Table 7.5 also shows that the coefficient of APF decreased after the implementation of the EPP. In the case of March and December, the APF coefficient dropped by about two-thirds after the introduction of the EPP. Figure 7.1 shows how the value of the APF coefficient changed during the entire sampling period. The linear approximation line clearly demonstrates that the value of the

Table 7.5 Results of hedonic price estimations (March, July, and December)

		2008	*2009*	*2010*	*2011*
Room Size	March		348.28*	1,024.80*	840.98*
	July	225.67*	616.62*	559.20*	
	December	465.29*	1,432.09*	1,248.40*	
APF	March		29,031.54*	24,030.77*	10,574.80*
	July	14,798.92*	16,859.37*	14,465.91*	
	December	28,850.04*	18,747.82*	8,806.41*	

Note: * indicates statistically significantly different from zero at the 5% level.

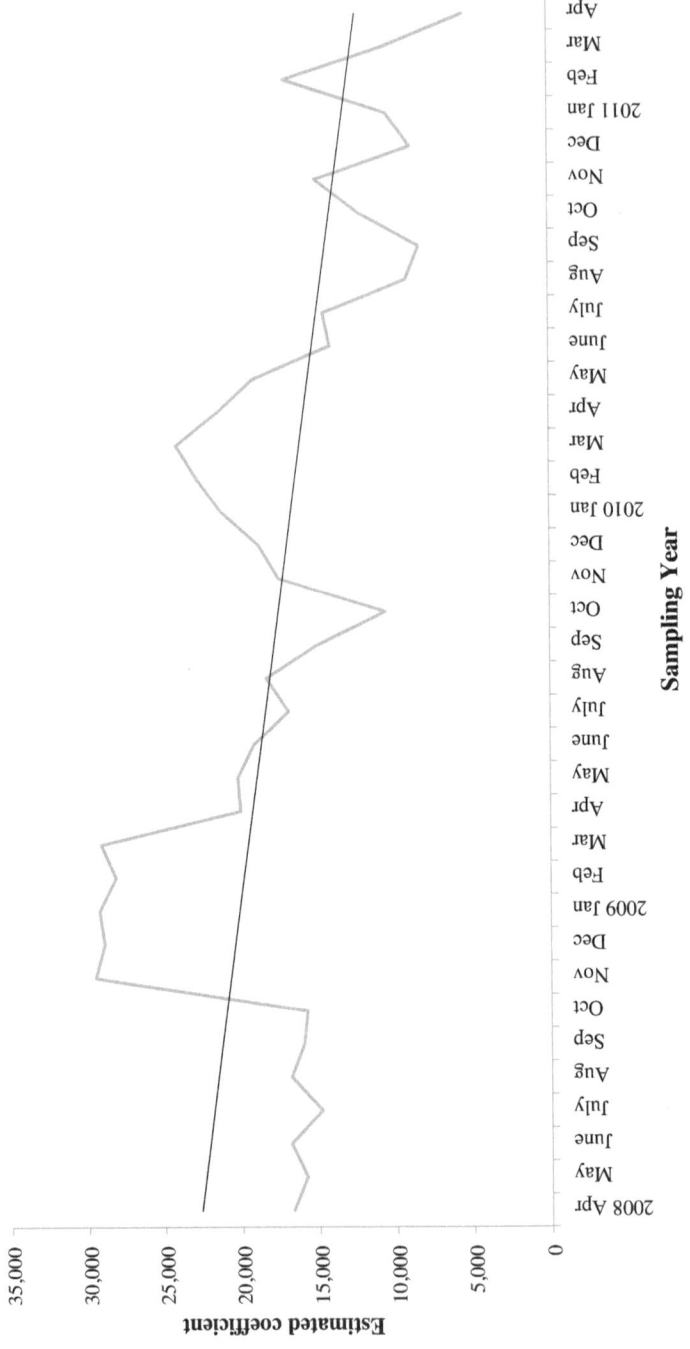

Figure 7.1 Consumer valuation of APF

APF coefficient declined after the introduction of the EPP. These results suggest that consumers placed lower value on energy saving and applied a higher implicit discount rate during the eco-point period. If the coefficient of APF is reduced to 10,000 in the previous example, then the implicit discount rate is estimated to be about 46 percent.

7.5. Conclusion

Many countries have started implementing rebate programs to promote energy-efficient electric appliances. In this chapter, we examined how the energy-efficient rebate program recently introduced by the Japanese government affected consumer valuation of energy efficiency.

We estimated the hedonic pricing model, which controls for the room size of air conditioners. The empirical results show that the energy-saving feature of air conditioners is reflected in their prices. However, the energy-efficient rebate program lowered consumer valuation of energy efficiency. Specifically, in the case of air conditioners designed for a 21 m² room, we found that the implicit discount rate increased from 26 percent to 46 percent during EPP.

The government can provide consumers with incentive to select products with a desirable attribute using a rebate program. However, the rebate program decreases the price of the attribute, and thus consumers place lower value on it. Only a small fraction of the rebate is spent for the targeted purpose.

Acknowledgments

An earlier version of this chapter was presented at the energy-saving mini-conference held at Nishiura City. The authors acknowledge valuable comments and suggestions from the participants.

Notes

1 Sanstad, Hanemann, and Auffhammer (2006) and Gillingham, Newell, and Palmer (2009) provide excellent survey information on the energy efficiency gap.
2 In 1992, the U.S. Environmental Protection Agency (EPA) introduced Energy Star as a voluntary labeling program designed to identify and promote energy-efficient products (U.S. EPA 2001). The label is now utilized in major global markets and helps consumers make an informed choice. It provides manufacturers with an incentive to develop more efficient products.
3 The effect of rebate programs has been intensively studied in recent literature. For example, Beresteanu and Li (2011), Chandra, Gulati, and Kandlikar (2010), and Gallagher and Muehlegger (2011) examined the impact of rebate programs for hybrid vehicles. However, to the best of our knowledge, no study has been done on the effect of rebate programs for home electric appliances.
4 The State Energy Rebate Program in the United States is a similar program that provides about $300 million to help consumers replace existing appliances with Energy Star–qualified appliances.

5 One eco point corresponds to 1 yen.
6 120 yen is converted into $1.
7 The residential electricity price in Japan is about 22.76 yen/kilowatt-hour, which is about twice of that of the United States (OECD/IEA 2010).
8 According to a survey by the Agency of Natural Resources and Energy of Japan (2004), air conditioners account for 25.2 percent of total household energy consumption.
9 We removed models with fewer than 500 sales.
10 Room sizes are also standardized in Japanese homes. The "Tatami mat" is used as a measurement unit for traditional Japanese rooms. The size of one Tatami mat is 1.74 meters by 0.87 meters. Room sizes are standardized according to the number of Tatami mats used. A salesperson in an electronics retail store initially asks customers about the size of the room in which an air conditioner is to be installed.
11 The Japanese fiscal year runs from April to March, so March is the peak period for moving.

References

Agency of Natural Resources and Energy of Japan. (2004) 'The summary of energy demand.' *White Paper of Energy*, Available: <http://www.enecho.meti.go.jp/about/whitepaper/2004html/2-1-2.html> (accessed 26 March 2015).

A. Beresteanu and S. Li. (2011) 'Gasoline prices, government support and the demand for hybrid vehicles in the United States', *International Economic Review* 52, 161–82.

A. Chandra, S. Gulati, and M. Kandlikar. (2010) 'Green drivers or free riders? An analysis of tax rebates for hybrid vehicles', *Journal of Environmental Economics and Management* 60, 78–93.

M.K. Dreyfus and K.W. Viscusi. (1995) 'Rates of time preference and consumer valuations of automobile safety and fuel efficiency', *Journal of Law and Economics* 38, 79–105.

K.S. Gallagher and E. Muehlegger. (2011) 'Giving green to get green? Incentives and consumer adoption of hybrid vehicle technology'. *Journal of Environmental Economics and Management* 61, 1–15.

K. Gillingham, R.G. Newell, and K. Palmer, (2009) 'Energy efficiency economics and policy', *Discussion paper: Resource for the Future*. RFF DP 09–13.

L.A. Greening, A.H. Sanstad, and J.E. McMahon. (1997) 'Effects of appliance standards on product price and attributes: an hedonic pricing model', *Journal of Regulatory Economics* 11, 181–94.

J.A. Hausman. (1979). 'Individual discount rates and the purchase and utilization of energy-using durables', *Bell Journal of Economics* 10, 33–54.

R.B. Howarth and A.H. Sanstad, (1995) 'Discount rates and energy efficiency', *Contemporary Economic Policy* 13, 101–9.

O. Kimura. (2010) 'Japanese Top Runner approach for energy efficiency standard', *Discussion Paper: Socio-Economic Research Center*. Central Research Institute of Electric Power Industry. SERC09035.

Ministry of Economy, Trade and Industry. (2011) 'Policy Impact of eco-point program', Available: <http://www.meti.go.jp/press/2011/06/20110614002/20110614002-2.pdf> (accessed 10 July 2011) (In Japanese).

OECD/IEA. (2010) *Energy Prices and Taxes*. Vol. 2010/1, OECD Publishing, Paris.

D. Revelt and K. Train. (1998) 'Mixed logit with repeated choices: Households' choices of appliance efficiency level', *Review of Economics and Statistics* 80, 647–57.

S. Rosen. (1974) 'Hedonic prices and implicit markets: product differentiation in pure competition', *Journal of Political Economy* 82, 34–55.

H. Ruderman, M.D. Levine, and J.E. McMahon. (1987) 'The behavior of the market for energy efficiency in residential appliances including heating and cooling equipment', *Energy Journal* 8, 101–14.

A.H. Sanstad, W.M. Hanemann, and M. Auffhammer. (2006). 'End-use energy efficiency in a "post-carbon" California economy: policy issues and research frontiers', In M.W. Hanemann and A.E. Farrell (Eds.), *Managing Greenhouse Gas Emissions in California*, Chapter 6. The California Climate Change Center at UC Berkeley.

U.S. Energy Information Administration. 2011. *International Energy Outlook 2011.* Washington, D.C., U.S. Department of Energy.

U.S. Environmental Protection Agency. (2011) 'Current and near-term greenhouse gas reduction initiatives', Available: <http://www.epa.gov/climatechange/policy/neartermghgreduction.html> (accessed 20 August 2011).

Part III

Rebate program for eco-friendly vehicles

The Japanese motor vehicle market

Shigeru Matsumoto

1. The motor vehicle market in Japan

1.1. Motor vehicle production

The motor vehicle industry is one of the most important industries in Japan. Automobile manufacturers employed 785,000 workers in 2013 (Statistical Bureau, Ministry of Internal Affairs and Communications 2013), accounting for 8.7 percent of the total labor force (Japan Automobile Manufacturers Association [JAMA] 2014a). The total value of vehicle shipments from automobile manufacturers was 52 trillion yen ($418 billion) in 2012,[1] which was 17.4 percent of the total shipment value of the entire manufacturing sector (Ministry of Economy, Trade and Industry [METI] 2012).

Figure III.1 shows changes in the production, registration, and export of Japanese passenger vehicles over time. The figure demonstrates that production increased steadily from 1995 to 2008 while the number of new vehicle registrations remained roughly the same. Consequently, the export/production ratio has increased since 2000, reaching its peak of 59.6 percent in 2008.

After the Lehman Brothers bankruptcy, production dropped by about 70 percent. To stimulate the local economy, the Japanese government introduced the first eco-car subsidy program, from June 2009 to September 2010. Owing to this program, about one-third of the production loss caused by the Lehman Brothers bankruptcy was recovered. However, in 2011, the recovered production was immediately lost. The reduction in production was roughly the same as the reduction in vehicle registrations. To further stimulate the local economy, the Japanese government implemented a second eco-car subsidy program, from April 2012 to September 2012. After this second subsidy program, production went back up to the level of the early 2000s.

Figures III.2a and III.2b show the production shares of the Japanese automobile manufacturers. Nine manufacturers produced 8,493,943 passenger vehicles in 1993, while eight manufactures produced 8,189,323 passenger vehicles in 2013. Toyota has maintained the largest production share for the last 20 years and had 37 percent of the production share in 2013. Conversely, Nissan, Mitsubishi, and Honda have lost production share during the last 20 years, while other manufacturers have expanded their share.

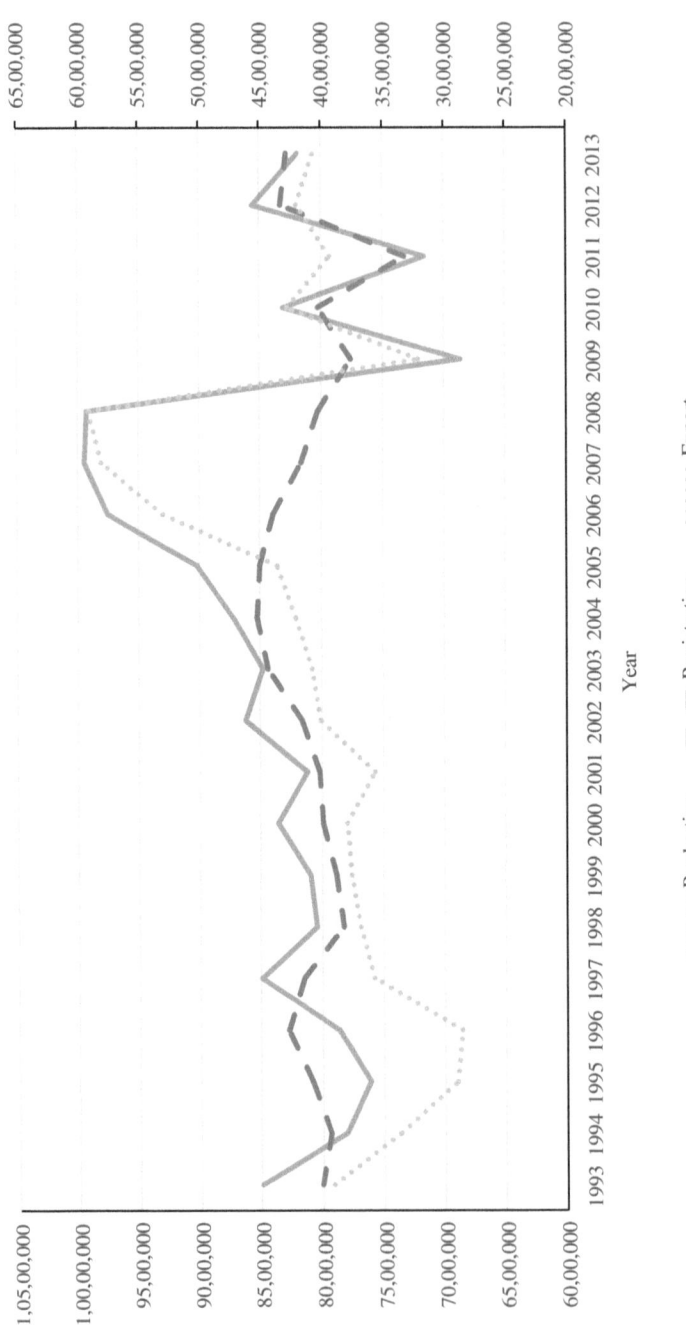

Figure III.1 Change in the production/registration/export of passenger vehicles
Source: JAMA (2014b).

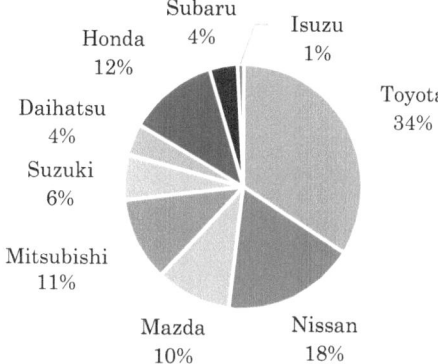

Figure III.2a Production share of passenger vehicles (1993)
Source: JAMA (2014b).

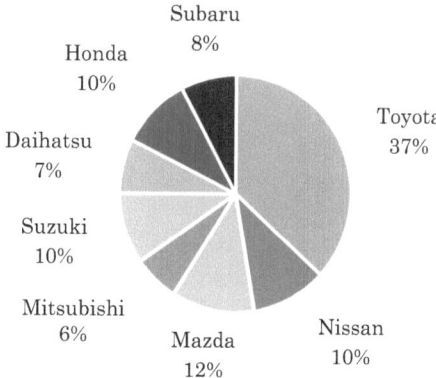

Figure III.2b Production share of passenger vehicles (2013)
Source: JAMA (2014b).

1.2. Motor vehicle sales

In 2013, the total number of new passenger vehicles sold in the Japanese market was 4,562,184. The market share of regular cars, compact cars, and light motor vehicles were 0.31, 0.32, and 0.37 percent, respectively.[2] Figure III.4 compares the change in market share across three vehicle size classes. The figure shows that market shares of regular and compact cars have decreased for the last several years, while those of light motor vehicles have increased. This suggests that many Japanese have downsized their vehicles.

According to 2013 registration data, the market share of used regular car registration as a portion of total regular car registration was 54.36 percent,

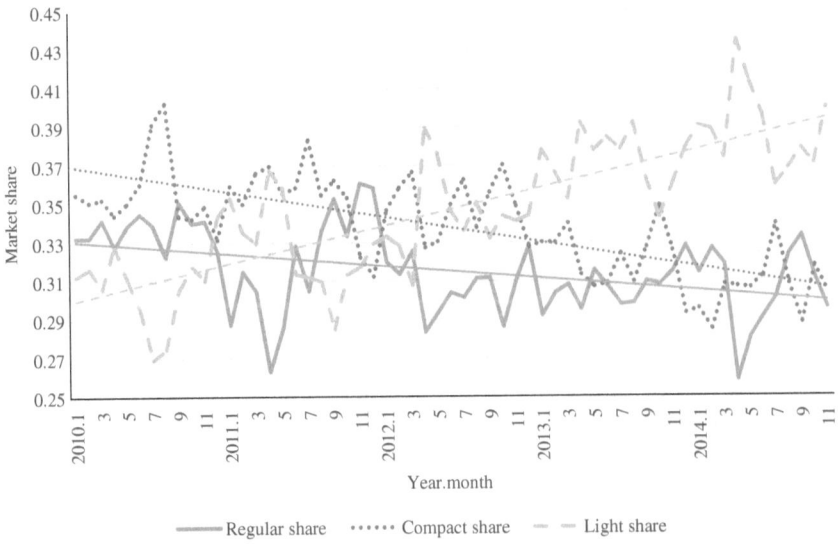

Figure III.3 Vehicle size and market share

Source: JAMA (2014a).

while that of used compact car registration among total compact car registration was 54.17 percent (JAMA 2014a). Except for 2011 (the year of the Great Tohoku Earthquake), the market share of used cars has remained at around 40 to 50 percent.

Sales of foreign-brand imported vehicles in the Japanese market declined from 340,000 in 1996 to 160,000 in 2006. This trend has reversed since 2007, and sales rose to 280,000 in 2013.[3] These sales accounted for 8.6 percent of the total sales of new vehicles (Japan Automobile Importers Association 2014). Among foreign-brand imported vehicles, German cars are popular in the Japanese market. Volkswagen had the largest market share of foreign-brand vehicles (19.93 percent) in 2013, while Mercedes-Benz was second (16.51 percent) and BMW was third (13.88 percent) (Nikkei Newspaper 2014).

1.3. Motor vehicle use

Figure III.4 presents the change in the number of registered passenger vehicles. This number increased rapidly from 1990 to 2000. However, the growth rate has declined since 2000. This implies that the domestic market has been saturated due to population decreases and an aging economy. The figure further demonstrates that the market share of compact cars has decreased since 1992, while that of both regular cars and light motor vehicles has increased.

Figure III.5a compares the number of registered vehicles per household across prefectures. Not surprisingly, the number is small in metropolitan areas. Households in the Tokyo, Kanagawa, Osaka, Kyoto, and Hyogo prefectures have no more than

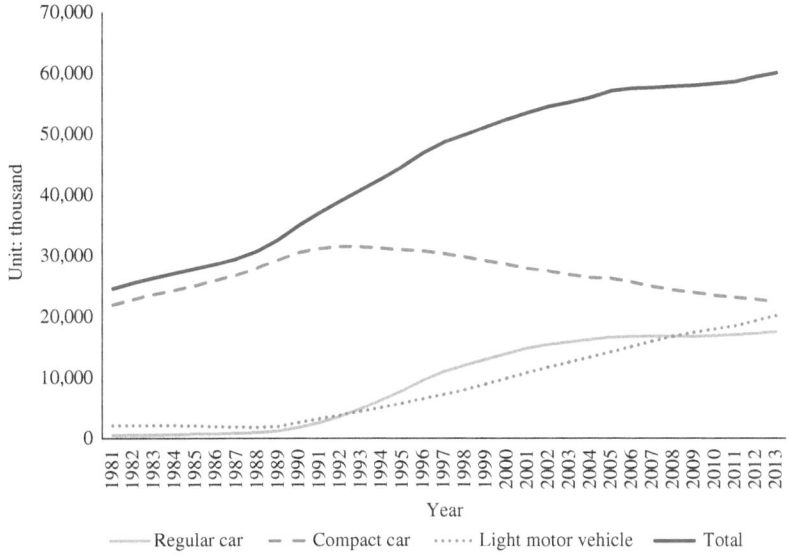

Figure III.4 Change in the number of registered passenger vehicles in Japan
Source: MLIT (2014a).

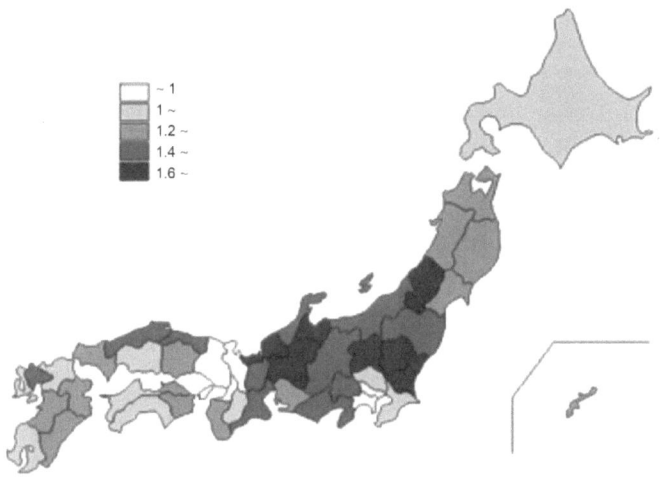

Figure III.5a Number of registered vehicles per household
Source: Automobile Inspection and Registration Information Association (2013).

one vehicle, while households in other prefectures have more than one vehicle. The
Ministry of Land, Infrastructure, Transport, and Tourism (MLIT 2014a) estimated
the annual household gasoline consumption in each prefecture. We divided this
number by the prefecture population and calculated the average annual gasoline
consumption per capita. According to our calculation, the average Japanese

person consumes 365.55 liters of gasoline and emits 862.698 kg of CO_2 annually.[4] Further, the pattern of gasoline consumption presented in Figure III.5b is very similar to the pattern of vehicle ownership presented in Figure III.5a. This implies that people in metropolitan areas consume less gasoline.

MLIT (2014a) estimated the annual travel distance of passenger vehicles in each prefecture. We divided this number by the prefecture population and calculated the average annual travel distance per capita. According to our calculation,

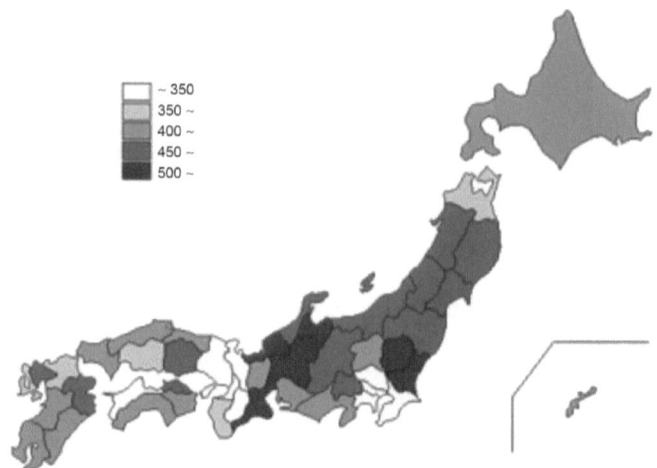

Figure III.5b Average annual gasoline consumption per capita (kl)
Source: MLIT (2014a).

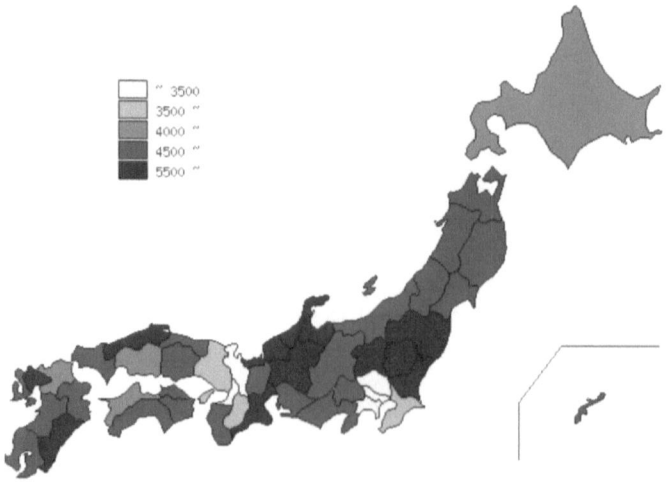

Figure III.5c Distance traveled per person (km)
Source: MLIT (2014a).

the average Japanese person travels 4,065.89 km per year. Figure III.5c compares travel distance across prefectures, and it shows that people in metropolitan areas travel less by car.

2. Fuel economy standards

The Japanese government introduced the Act Concerning the Rational Use of Energy (Energy Conservation Act) in 1979 and specified the machineries and appliances in particular, that required improvement in energy efficiency on the usage. Both passenger vehicles and trucks are specified as specific energy consumption apparatus. Under this act, automobile manufacturers and importers that sell vehicles in Japan are requested to achieve the fuel economy standard prior to the target year.[5]

In 1998, METI established the Top Runner program. This program designates the most fuel-efficient vehicle available on the market in each weight class as a "top runner" and sets the fuel economy of the "top runner" as the fuel economy standard. In 2015, vehicles will be classified into 16 weight classes. Table III.1

Table III.1 Fuel economy standard (gasoline passenger vehicles for 2015)

Weight (kg)	Fuel economy standard (km/l)
0–601	22.5
602–740	21.8
741–855	21.0
856–970	20.8
971–1,080	20.5
1,081–1,195	18.7
1,196–1,310	17.2
1,311–1,420	15.8
1,421–1,530	14.4
1,531–1,650	13.2
1,651–1,760	12.2
1,761–1,870	11.1
1,871–1,990	10.2
1,991–2,100	9.4
2,101–2,270	8.7
$\geq 2,271$	7.4

Source: MLIT (2014b).

presents the fuel economy standard for 2015. As the table shows, a less stringent standard is applied for heavy vehicles.

The fuel economy standard has been updated several times. Figure III.6 shows the historical change in average fuel economy, based on the 10.15 mode. The 10.15 mode is the load-testing procedure used to measure fuel economy in Japan (see further explanation about the 10.15 mode in Chapter 9 on page 209). The figure also shows that fuel economy has been improved in all weight classes. In fact, for the last 15 years fuel economy has increased by about 40 to 50 percent.

It is well known that there is a wide discrepancy between catalog and real fuel economy. We estimated the real fuel economy of passenger vehicles based on the dataset from MLIT (2014b). According to our estimation, the real fuel economies of regular, compact, hybrid, and light passenger vehicles are 8.36, 11.28, 16.03, and 13.32 km/l, respectively. Therefore, CO_2 emissions of regular, compact, hybrid, and light passenger vehicles are 275.12, 203.90, 143.48, and 172.67 g/km, respectively. These values are much higher than the ones based on the 10.15 mode estimation. For instance, in our calculation, although the CO_2 emission of an average hybrid vehicle (HV) is 143.48 g/km, the Toyota Prius is assumed to emit only 61 kg/km on the 10.14 mode estimation (MLIT 2014b).

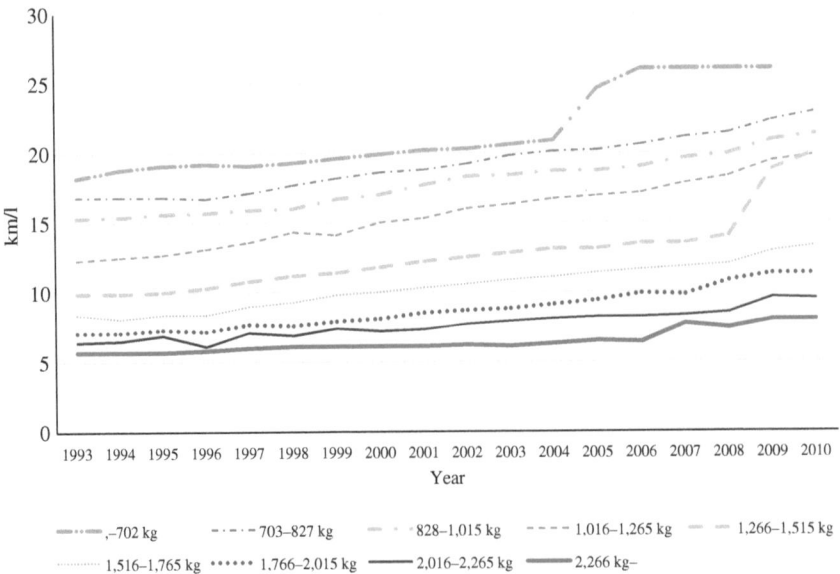

Figure III.6 Change in fuel economy (10.15 mode)
Source: MLIT (2014b).

3. Policies to promote next-generation vehicles

The energy efficiency of the transportation sector will be improved substantially through the market penetration of next-generation vehicles. For the last several years, both the national and local governments have implemented various policy measures to promote next-generation vehicles. In 2014, the governments provided subsidies to buyers of plug-in hybrid vehicles (PHVs) and electric vehicles (EVs). The amount of the subsidy varies according to the price of the vehicle. For instance, the amount of the subsidy for the Toyota Prius is 330,000 yen, while that for the Nissan Leaf is 520,000 yen (Next Generation Vehicle Promotion Center, 2014). In addition to the subsidy for the vehicle purchase, the national and local governments have provided subsidies for charging facilities for PHVs and EVs. On average, half of the construction fees for charging stations have been subsidized. Further, the governments provide preferential tax treatments to eco-friendly vehicles.

In 2013, the total number of "regular" HV sales in the Japanese market was 1,011,081 (Next Generation Vehicle Promotion Center, 2014), while the total number of passenger vehicle sales was 4,283,338 (JAMA 2014b). Therefore, HV sales accounted for 23.61 percent of total passenger vehicle sales in 2013. In the same year, the number of registered HVs was 3,793,075, and the total number of registered passenger vehicles was 60,035,000. Therefore, the share of HVs among all passenger vehicles was about 6.32 percent. These statistics confirm that HVs have already been popular vehicles among the Japanese. In contrast, the market penetration of PHVs and EVs has been much slower. In 2013, the number of EVs and PHVs sold in the Japanese market were 16,818 and 12,972, respectively. The number of registered EVs was 54,664, while that of PHVs was 30,171. Thus, additional support programs are required for the promotion of EVs and PHVs.

Notes

1   120 yen converts to $1.
2   Vehicles with engines 660cc or under are classified as light motor vehicles.
3   This number does not include reverse-imported vehicles.
4   We assume that 1 liter of gasoline emits 2.3 kg of CO_2.
5   Manufacturers have to make their production-weighted harmonic mean fuel economy surpass the fuel economy standard.

References

Automobile Inspection and Registration Information. (2013) 'Data of the number of vehicles owned', Available: <http://www.airia.or.jp/number/index.html> (accessed 16 December 2014).

Japan Automobile Importers Association. (2014) 'Imported car market of Japan', Available: <http://www.jaia-jp.org/j/data/market/> (accessed 10 December 2014).

Japan Automobile Manufacturers Association, Inc. (2014a) 'Japan auto industry', Available: <http://www.jama.or.jp/industry/index.html> (accessed 10 December 2014).

Japan Automobile Manufacturers Association, Inc. (2014b) 'Japan auto industry data', Available: <http://jama.org/japan-auto-industry-data/database> (accessed 8 December 2014).

Ministry of Economy, Trade and Industry. (2012) 'Census of manufacturers', Available: <http://www.meti.go.jp/english/statistics/tyo/kougyo/index.html> (accessed 12 December 2014).

Ministry of Land, Infrastructure, Transport and Tourism. (2014a) 'Survey of vehicle energy consumption', Available: <http://www.mlit.go.jp/k-toukei/nenryou/nenryou.html> (accessed 12 December 2014).

Ministry of Land, Infrastructure, Transport and Tourism. (2014b) 'About the fuel economy target', Available: <http://www.mlit.go.jp/jidosha/jidosha_fr10_000005.html> (accessed 15 December 2014).

Next Generation Vehicle Promotion Center. (2014) 'Subsidy information', Available: <http://www.cev-pc.or.jp/english/index.html> (accessed 14 December 2014).

Nikkei Newspaper. (2014, 4 April) '2013 imported car market share', Available: <http://www.nikkei.com/article/DGXNASFL030OM_T00C14A4000000/> (accessed 8 December 2014).

Statistical Bureau, Ministry of Internal Affairs and Communications. (2013) 'Labour force survey', Available: <http://www.stat.go.jp/data/roudou/2.htm> (accessed 18 December 2014).

8 Cross-country policy comparison

What are the elements of a successful eco-car policy?

Keiko Hirota and Shigeru Kashima

8.1. Introduction

There is often a delay in drafting and implementing policies for emissions/CO_2 reduction and energy saving in developing countries. One of the reasons for such delays is that the procedure for drafting such policies is often not clearly understood. Policies need to be tailor-made, and policy makers often struggle to devise effective policies. Thus, the transfer of knowledge on how to devise and implement policies for emissions/CO_2 reduction from countries that have already introduced them becomes crucial.

This chapter aims to describe the general trend of policy implementation using examples of policies for eco-friendly new passenger vehicles (hereafter, "eco-car policies") and the corresponding market responses in eight emerging economies: Brazil, China, India, Indonesia, Malaysia, the Philippines, South Africa, and Thailand. Eco-car policies may work through a combination of technical requirements and economic incentives. Today, eco-car policies are widely used but in a limited manner. This chapter also attempts to evaluate eco-car policies by comparing the acquisition costs of five passenger cars (new cars) for each country. The comparative study of this policy area provides information about the additional reforms needed across the entire vehicular system. Thus, this chapter hopes to answer the question, "what are the elements of a successful eco-car policy?"

8.2. Common tools for eco-car policies

8.2.1. Technological and economic policies

This section compares the following five technical measures for eco-car policy implementation among the aforementioned eight emerging economies/developing countries and Japan: (1) vehicle registration, (2) emissions standard, (3) fuel economy standard, (4) introduction of hybrid/electric/fuel cell vehicles (referred together as xEVs in this chapter) in the market, and (5) mandatory inspection (Table 8.1). These measures play important roles in devising incentives for eco-cars. Like Japan, all eight developing countries are already implementing emission standards for new passenger cars.

Table 8.1 Policy and measures of eco-cars (passenger cars) by country (2014)

● The measure in operation

Policy for technical performance in a vehicle	Measure	New car/ car in-use	Brazil	China	India	Indonesia	Malaysia	Philippines	South Africa	Thailand	Japan
Tailpipe emission reduction	Emission standard	New car	● Proconve 5 (EURO4 equivalent)	● China 4 (equivalent)	● Bharat 3 (equivalent)	● EURO 2	● EURO 3 (Gasoline) EURO2 (Diesel)	● EURO 2	● EURO 2	● EURO 4	● 2009 Post new long term
Fuel consumption reduction / fuel economy improvement/ CO2 reduction	Fuel economy/ consumption/ efficiency standard	New car	●	●	● (Forthcoming)	No standard	No standard	Real world fuel economy measurement	No standard	● (Forthcoming)	●
New energy vehicle, New car replacement	Introduction xEVs (EV, HEV, FCEV) in the market	New car	● Fuel Flex Vehicle (FFV), EV	● EV, HEV FCEV (Project only)	● EV (Under development)	● EV (Under development)	● EV, HEV	● E tricycle (Project only)	● EV, HEV	● EV, HEV	● EV, HEV FCEV (Forthcoming)
Technical control	Mandatory inspection	Car in-use	● Every year	● Vehicle age 4, 6, 8, 10, 12 and 14. Every year after vehicle age 14.	● Vehicle age 1, 15 and 20.	No inspection	No inspection	● Every year	No inspection	● Every year after vehicle age 7	● Vehicle age 3 Every 2 years after vehicle age 3

Prepared by the author.

The uptake of low-emission cars depends on the availability of fuel quality appropriate to the emission standard, which, in turn, requires updating fuel refinery and distribution infrastructure. China and South Africa have already implemented fuel economy standards, while Thailand and India plan to in the near future. The rest of the countries that are part of this study do not have any current plans in this regard. According to the available evidence, the applied standards and categories of this market segment by weight, vehicle type, fuel type, and sales volume are country specific. Moreover, before they are passed into law, all standards for fuel economy need the approval of various stakeholders, such as the government, automobile associations, academia, and consumers. It is notable that the considerable efforts of various Japanese stakeholders have helped Japan to successfully implement its fuel economy policy earlier compared to other countries. While xEVs are already available for sale in the abovementioned eight countries, their numbers are still limited (as indicated by the number of vehicles registered); the prices of xEVs and the availability of related infrastructure are key for their dissemination. As of 2015, fuel cell electric vehicles (FCEVs) will available for purchase in Japan. The establishment of hydrogen fueling stations is a key issue for the dissemination of FCEVs in Japan. Five of the eight developing countries conduct mandatory inspection for passenger cars alongside mandatory inspection for commercial vehicles. The implementation of mandatory inspection for passenger cars depends on the capacities of the inspection institutes (namely, the number of inspection lanes, availability of measuring equipment, and number of inspectors). In Japan, mandatory inspection is an opportunity for car owners to replace their older/less efficient cars with newer/more efficient vehicles. This may be why the average vehicle age is lower in Japan compared to that in developing countries. This section compares the technical abilities of the eight countries and Japan to help clarify realistic procedures for eco-car policy implementation in developing countries.

8.2.2. Vehicle registration

The number of vehicles registered is a basic indicator to understand vehicle ownership in each country. Data availability in this case depends on specific data items (vehicle type, vehicle age, fuel type, etc.) and frequency of updates (registration renewal and deletion). The number of vehicles registered can provide important inputs to other policies (such as tax collection, recall, traffic control, crime prevention, curbing illegal export, proper dismantling, etc.) and timings for announcements of mandatory inspection. These statistics can also be used to estimate emissions of CO_2 and other air pollutants. These statistics may enable us to measure the quantitative impact of policy implementation in terms of the emission standard, fuel economy standard, and xEV type.

Data availability by number of registered vehicles and types of registration systems is shown in Table 8.2. Data for the number of vehicles owned by vehicle type (passenger cars, commercial vehicles, and motorcycles) are available for all eight countries and Japan. Registration by vehicle age is available to the public only in Thailand and Japan. Registration by fuel type is a basic indicator of fuel consumption and fuel

Table 8.2 Availability of statistical data (2014)

		Brazil	PR China	India	Indonesia	Malaysia	Philippines	Thailand	South Africa	Japan
Data availability	by vehicle type	●	●	●	●	●	●	●	●	●
	by vehicle age	No	No	No	No	No	No	●	No	●
	by fuel type	●	● (Sales only)	No	No	No	●	●	No	●
	New registration (Sales)	●	●	●	●	●	●	●	●	●
	Annual data renewal	●	●	Vehicle age 15 years or 8 years (*)	●	●	●	●	●	●
Registration system	Deletion procedure	NA	●	Automatic deletion at vehicle age 15 years or 8 years (*)	Automatic deletion after 2-3 years in case of no-annual tax payment (**)				NA	●

● The measure in operation
No: There is no measure.
NA : No answer

* Case of Delhi and Mumbai. The vehicle age is 15 years for vehicles for personal use and 8 years for commercial vehicles.
** Comments from vehicle experts vehicle-related institutes.

Data sources: ANFAVEA (2013), Hirota (2012), and NAAMSA (2014).

Prepared by the author.

quality policy. Four out of the eight countries collect registration data by fuel type, as does Japan. All eight countries record all new registrations (sales). The automotive associations in the eight countries collect the sales data from association members. These data are available on their respective websites.

Renewal and deletion procedures enable the Japanese and Chinese authorities to update vehicle registration data. Indonesia, Malaysia, the Philippines, and Thailand automatically delete registrations after two or three years unless a payment for renewal is made. All eight countries except India require renewal via an annual tax payment.

In Delhi and Mumbai (India), the registrations of personal vehicles completing 15 years are automatically deleted. Personal vehicles aged 15 years or more can be registered again after passing an automobile inspection. This procedure may force older vehicles to be phased out. Procedures are in place to delete a registration if a vehicle is stolen, unfit for service, or involved in an accident. However, these procedures are not obligatory. Consequently, there may be significant discrepancies between the registered ownership status and the actual status by use. Without a deletion procedure, the number of registrations may exceed the actual number of vehicles on the road.

A typical data collection method for vehicles can be explained using the example of India. Vehicle registration is based on the Motor Vehicles Act (1988) and the Central Motor Vehicles Rules (1989). Currently, the data are collected by various organizations involved in the data registration system. Owners are required to register their new vehicles at the time of purchase. The data items recorded include registration place and vehicle type, namely its weight, fuel capacity, engine capacity, and number of seats. Maharashtra, an Indian state, publishes a statistical yearbook annually, which includes the number of registrations by region and vehicle type. The Transport Research Wing (TRW) of the Indian Government operates in all the states (e.g. there are 13 centers in Delhi and 46 in Mumbai). The data collected by the registration centers are first sent to the Transport Commissioner's Office of each state and are then transmitted to TRW. Thereafter, the data are sent to the statistics department of India's Ministry of Shipping, Road Transport and Highways, which publishes the national statistics on an annual basis.

While the *Road Transport Yearbook* includes the total number of registered vehicles, the number of newly registered vehicles is not mentioned. Table 8.3 shows the number of total registered vehicles by state from 2000–6 (TRW 2009). It is clear that data availability differs by state. Therefore, it is desirable that a common timeline be followed for recording all automotive statistical data throughout a country in order to understand trends, conduct estimations, and assess policy implementation successfully.

8.2.3. Emission standards

The introduction of emission standards intended to limit harmful emissions from new vehicles is the first step toward establishing an eco-car policy. Table 8.4 shows the time schedules of the introductions of vehicle emission standards for

Table 8.3 Total registered vehicles by state (India, 2009)

(**In thousands**)

STATES/UTs	2000	2001	2002	2003	2004	2005	2006(P)
1	2	3	4	5	6	7	8
STATES							
Andhra Pradesh	3636	3966	4389	5002	5720	6458	7218
Arunachal Pradesh	21*	21*	21*	21*	21*	22	22
Assam	453	542	596	657	727	815	914
Bihar	871#	949	1024	1121	751	1352	1432
Chhattisgarh	–	857	948	1076	1216	1375	1541
Goa	319	341	366	397	436	482	529
Gujarat	5189	5576	6008	6508	7087	7817	8622
Haryana	1733	1949	2122	2279	2548	2854	3087
Himachal Pradesh	193	217	244	269	289	301	334
Jammu & Kashmir	299	330	364	399	439	478	524
Jharkhand	–	909	984	1101	1217	1357	1505
Karnataka	3393	3537	3636	3738	3977	5436	6220
Kerala	1782	2112	2315	2552	2792	3122	3559
Madhya Pradesh	3457	3095	3173	3459	3804	4188	4609
Maharashtra	6114	6760	7414	8134	8969	9936	10966
Manipur	77	77$	90	97	106	114	124
Meghalaya	58	62	67	73	73^	92	104
Mizoram	27	31	34	37	42	47	52
Nagaland	145	160	177	162	172	172	184
Orissa	982	1096	1215	1359	1525	1715	1932
Punjab	2296**	2910	3103	3308	3529	3876	4035
Rajasthan	2712	2943	3197	3487	3834	4261	4754
Sikkim	12	12	13	15	17	20	22
Tamil Nadu	4611	5162	5658	8005	8575	9257	10054
Tripura	45	50	57	66	76	73	106
Uttarakhand	–	364	406	457	516	573	643
Uttar Pradesh	4627	4921	5171	5928	6460	7344	7989
West Bengal	1690	1690$	1690$	2366	2548$	2681	2872
TOTAL STATES	**44742**	**50639**	**54482**	**62073**	**67466**	**76218**	**83953**
UTs							
A & N Islands	23*	25	28	28+	28+	37	41
Chandigarh	386**	386**	386**	562	586	617	647
D & N Haveli	13*	13*	13*	31	35	40	45
Daman & Diu	34	37	41	44	48	51	55

Delhi	3424	3635	3699	3971	4237	4187	4487
Lakshadweep	4	4	5	5	5	5	6
Puducherry	231	252	270	293	313	347	384
TOTAL UTs	**4115**	**4352**	**4442**	**4934**	**5252**	**5283**	**5665**
GRAND TOTAL	48857	54991	58924	67007	72718	81502	89618

* Data relates to 1996–97.
** Data relates to 1997–98.
$ Data relates to 1999–00.
+ Data relates to 2001–02.
^ Data relates to 2002–03.
# Figures for Bihar only, excluding Jharkhand.
(P) Provisional
Note: Consequent on formation of three States (Chhattisgarh, Jharkhand and Uttarakhand), separate data is available only from 2000–01.
Data source: TRW (2009).

passenger cars (gasoline and diesel) in the eight countries and Japan. The introduction of the emission standard schedule should be coupled with the required changes in fuel quality and vehicle equipment modifications. For example, leaded gasoline should have been phased out before the EURO 1 standard was introduced, because lead in gasoline causes the catalytic converter to malfunction. Sulfur content in fuel should have been lowered to below 500 ppm before the introduction of the EURO 2 standard and below 150 ppm and 350 ppm for gasoline and diesel vehicles, respectively, before introduction of the EURO 3 standard. Similarly, sulfur content in fuel should have been limited to below 50 ppm before the introduction of the EURO 4 standard. Catalytic converters were required to be installed to meet the EURO 2–mandated emission levels for CO, HC, NOx, etc. The combined efforts of the government, automobile manufacturers, and the fuel industry ensured that Japan introduced stringent emission standards earlier than the eight developing countries. Distribution of fuel of an appropriate quality is a common issue faced by developing countries with regard to emission standards. Large countries, such as China and India, started by introducing more stringent emission standards at the city level. However, Brazil implements emission standards on a nationwide basis.

The time schedule for implementing these changes seems to be very tight from the viewpoint of fuel refinery upgrades. In general, five years of lead time is required to update refinery technology to supply fuel of a quality meeting the EURO 4 standard. All of China and Brazil have managed to achieve this requirement, hence allowing them to progress from the EURO 3 to the EURO 4 standard in a comparatively short period. Initially, Indonesia hoped to transition directly from the EURO 2 standard to the EURO 4 standard, but the Ministry of Environment intervened, and Indonesia is now likely to introduce the EURO 3 and EURO 4 standards in 2015 and 2020, respectively. The time schedule for the introduction of the standards differs for gasoline and diesel vehicles in Malaysia. Malaysia introduced the EURO 3 standard for gasoline passenger cars and the EURO 2 standard

Table 8.4 Time schedule for passenger car emission standard of the eight countries 1996–2022 (fuel type, vehicle category, area)

Country	Fuel type	Vehicle category	Area	1996	1997	1998	1999	2000	2001	2002	2003	2004	2005	2006	2007	2008	2009	2010	2011	2012	2013	2014	2015	2016	2017	2018	2019	2020	2021	2022
Brazil	G (PC), G,D (LCV)	PC, LCV	Nationwide	Proconve 3 (EURO 0/2)											Proconve 4 (Euro 3, US 94)		Proconve 5 (Euro 4, LEV)						Proconve 6 (Euro 5)					Proconve 7 (Proposal)		
	D	PC	Nationwide	Diesel passenger cars are not allowed in Brazil after 1970s.																										
PR China	G		Nationwide						China 1 (Euro 1)			China 2 (Euro 2)				China 3 (Euro 3)					China 4 (Euro 4)					China 5 (Euro 5)				
	D		Nationwide					China 1 (Euro 1)				China 2 (Euro 2)				China 3 (Euro 3)					China 4 (Euro 4)					China 5 (Euro 5)				
	G	M1*	Beijing							China 1 (Euro 1)		China 2 (Euro 2)		China 3 (Euro 3)		China 4 (Euro 4; (Euro 4 Police, Post (deliver, garbage collection, construction by local government))					China 5 (Euro 5)									
	D		Beijing					Diesel passenger cars are not allowed to register in Beijing for the moment.																						
	G,D		Shanghai					China 1 (Euro 1)			China 2 (Euro 2)						China 3 (Euro 3)				China 5 (Euro 5)									
India	G,D	M1*	Nationwide	1996 Norm		(1998 cat. converter norms)		Euro					Euro 2					Euro3												
			New Delhi	1996 Norm		(1998 cat. converter norms)		Euro	Euro 2				Euro 3					Euro 4												
			Metros (Mumbai, Kolkata & Chennai)	1996 Norm		(1998 cat. converter norms)		Euro 1		Euro 2			Euro 3					Euro 4												
			Other cities (Bangalore, Hyderabad, Ahmedabad and other 4)	1996 Norm		(1998 cat. converter norms)		Euro 1			Euro 2		Euro 3					Euro 4												
Indonesia	G,D	M1	Nationwide												Euro 2								Euro3 (plan)					Euro 4 (plam)		
Malaysia	G	M1	Nationwide					Euro 2													Euro3		Euro4 (plan)			Euro5 (plan)				
	D	M1	Nationwide		Euro 1															Euro2			Euro4 (plan)			Euro5 (plan)				
Philippines	G,D	M1*	Nationwide					Euro 1								Euro 2								Euro4 (plan)						
Thailand	G,D	M1	Nationwide					Euro 1						Euro 2			Euro 3				Euro 4									
South Africa	G,D	M1	Nationwide														Euro 2													
Japan	G	M1**	Nationwide	1978 Regulation				2000 New short term					2005 New long term				2009 Post new long term													
	D	M1**	Nationwide	1997/1998 Regulation						2002 New short term			2005 New long term				2009 Post new long term													

G: gasoline; D: diesel; PC: passenger car; LDV: light-duty vehicle; LCV: light-commercial vehicle

M1: Up to 9 passengers. Gross vehicle weight of up to 3,500 kg.

* Modified M1: Up to 6 passengers. Gross vehicle weight up to 2,500 kg.

** Modified M1: Up to 10 passengers.

The years of the first registration are shown in this table.

Data sources: Andrianto (2008), AAF (2008), CAI-Asia (2009), Chen (2009), Delphi (2014/2015), Ding, Y. (2009), Hirota (2008b, 2008d, 2009), ICCT/DieselNet (2014), Nacua and Mill (2010a), Prawiraatmadja (2008), Sengupta (2009), Wangwongwatana (2007), Yan (2010).

Prepared by the author.

for diesel passenger cars in 2012. It plans to introduce the EURO 4 standard in 2015, such that diesel vehicles would leapfrog directly to the EURO 4 standard.

In 1999, the government of Beijing began incentivizing the replacement of old cars by implementing a labeling system by emission level. Gasoline passenger cars meeting the national emission standards of China 1 and diesel passenger cars meeting the national emission standards of China 1 and China 2 are certified as "yellow-label" vehicles. The remaining vehicles that emit less pollutants (gasoline vehicles following the China 2, China 3, and China 4 standards and diesel vehicles following the China 3 and China 4 standards) are certified as "green-label" vehicles. The government provides subsidies for replacing yellow-label vehicles. This incentive resulted in the phasing out of 140,000 yellow-label vehicles and reducing an estimated 200 tons of daily emissions from these vehicles.

8.2.4. Fuel economy standards

According to a report of the Global Fuel Economy Initiative (GFEI 2014), the average global fuel economy of light-duty vehicles improved by 1.7 percent per year from 2005–8. The pace of improvement accelerated slightly to 1.8 percent per year from 2008–11. The trend is encouraging, but more remains to be achieved; many developing countries do not have a fuel economy regulation yet. This section reviews developments in the fuel economy policies of Brazil, China, India, Thailand, the Philippines, and Japan. Table 8.5 shows fuel economy regulations by country. This comparative study shows a great deal of diversity in the design of the standards.

The Brazilian government implemented a program called "Inovar-Auto" to encourage automakers to produce efficient vehicles domestically. Inovar-Auto offers incentives in two ways (Presidência da República, 2012). First, it increases a tax on industrialized products (IPI) by 30 percent for all light-duty and light-commercial vehicles. Second, it imposes a series of requirements (concerning vehicle efficiency, domestic production, and investment in R&D and automotive technology) for automakers to qualify for an up to 30 percent discount in the IPI. Thus, the IPI tax remains unchanged for manufacturers that meet the abovementioned requirements. The program expects to reduce fuel consumption by 12 to 19 percent from 2013–17.

China issued the first mandatory national-level fuel economy regulation in 2004. The regulation is primarily designed to mitigate oil dependence. Its other objectives include encouraging domestic car manufacturers to import efficient vehicle technologies into the Chinese market and to phase out inefficient models (Wagner et al. 2009). The regulation has been implemented two phases. Phase 1 was introduced from July 2005 for newly approved models and from July 2006 for already introduced models. Phase 2 was introduced from January 2008 for newly approved models and from January 2009 for already produced models. Phase 3 was introduced in 2012 and is expected to achieve all the planned milestones by 2015. The Chinese government also plans to replace the average fuel economy standard with a corporate average fuel economy standard (CAFE) in Phase 3.

China's fuel economy regulation classifies vehicles into 16 categories by weight. This is similar to the Japanese system except that the Japanese system has fewer categories. The Chinese car market is more diverse and includes more than 100 car

Table 8.5 Fuel economy regulation by country 2015–21

	Target year	Standard type	Numerical target	Assessment	Test cycle	Tax incentive	Mandatory/ voluntary
Brazil	2017	Fuel economy	1.82 MJ/km	Weight-based corporate average	US Combined	Inovar-auto	Voluntary
China	2015	Fuel economy	6.9 liter /100 km	Weight class	EU NEDC		Mandatory
	2020		5 liter /100 km	Corporate average			
India (Proposed)	2016	CO2	130g / km	Weight-based corporate average	Indian NEDC		Mandatory
	2021		113 g / km				
Thailand (Proposed)	2016	Fuel economy	NA	Weight class	EU NEDC	Tax incentive for maximum limit	Minimum: Mandatory Maximum: voluntary
Japan	2015	Fuel economy	16.8 km / liter	Weight class	JC08	Eco-car tax	Mandatory
	2020		20.3 km / liter	Corporate average			

Data sources: DOE Philippines (2006), Gao and Jin (2011), GFEI (2014), Kovongpanich (2014), Tamang (2011), Thailand Automotive Institute (2014), and Wagner et al. (2009).

Prepared by the author.

manufacturers. Many of them supply only one or two models. A more fragmented categorization system could enable fairer assessments, which may help companies producing a limited number of models in the market to survive.

Another unique feature of China's fuel economy regulation is the introduction of the CAFE in Phase 3. The policy implementation of Phase 1 was assessed in 2008. The results indicated that the average fuel economy improved from 9.11 L per 100 km (2002) to 8.06 L per 100 km (2006), an improvement of 11.5 percent. The government plans to increase this rate to 20 percent (to about 7 L per 100 km) for Phase 3. The government has also set a target of 5 L per 100 km for fuel economy improvement by 2020. Implementation of the fuel economy regulation forced manufacturers to stop production of 444 types of vehicles because they did not meet the limits (Gao and Jin 2011). In total, it is estimated that the improvements in fuel economy have resulted in savings of more than 1 million tons of fuel.

In India, the Bureau of Energy Efficiency (BEE) hopes to finalize fuel efficiency standards for light-duty vehicles. It has proposed a fuel economy standard of 130 g per km by 2016 and 113 g per km by 2021. The proposed standard is represented as an upward sloping line relating fuel consumption level to the kerb weight of the vehicles. The draft is still being reviewed by the government, and the main discussion point is likely to center around the slope of the line for a particular target year.

Thailand's Department of Alternative Energy Development and Ministry of Energy ordered the Thailand Automotive Institute to prepare a draft fuel efficiency standard. The country follows two standards to promote high fuel efficiency in motor vehicles: the Minimum Energy Performance Standards (MEPS), which a vehicle must meet to be eligible for sale, and the High Energy Performance Standards, which stipulate the minimum energy efficiency requirements for any vehicle to avail tax promotion (Thailand Automotive Institute 2014).

Japan promotes clean energy vehicles (vehicles that emit less emissions and consume less energy). Fuel economy improvement is the most effective short-term strategy to reduce energy consumption and increase energy security. The Conference of Parties III (COP3) took place in Kyoto in December 1997. Developed countries agreed to cut greenhouse gas (GHG) emissions with specific targets. They were supposed to reduce GHG emissions by 5 percent relative to 1990 levels between 2008 and 2012. Japan had agreed to reduce GHG emissions by 6 percent relative to 1990 levels. Four main features of the fuel economy standard are considered as the key factors behind Japan's successful achievements in fuel economy improvements. First, the "Top Runner" program has helped Japan achieve its fuel economy targets. In the Top Runner program, the efficiency standard is set to that of the most efficient product currently available in the country's market. The Japanese government believed that the required reduction rate would be hard to achieve, but it was attainable because certain vehicle models had already achieved their required emission reduction targets. Second, Japan updated the fuel economy regulation regularly. The government introduced the mandatory fuel economy regulation with the Top Runner program for passenger vehicles and small trucks in March 1999. Japan planned to improve fuel economy by 22.8 percent from the 1995 level by 2010 (FE 2010). In fact, the fuel economy target was achieved in 2005, well ahead of schedule. In July 2007, Japan tightened fuel economy regulations for passenger vehicles, microbuses, small trucks, and mini-vans. Japan is expected to improve fuel economy for passenger vehicles by 23.5 percent from the 2004 level by 2015 (FE 2015). FE 2015 introduced a change in the driving cycle, from the 10/15 mode to the JC08 mode. The driving cycle of the JC08 mode is thought to be closer to real-world driving. In 2011, Japan set a new fuel economy target for 2020 (FE 2020). FE 2020 aimed to improve fuel economy by 24.1 percent of the corresponding value in 2009. Third, Japan's fuel economy regulation classifies vehicles into nine categories by weight, transmission, fuel, and vehicle type. It promotes advanced power train and vehicle technologies. Categorization tries to avoid unfair assessments between vehicles belonging to different groups (e.g. when there is a shift from a lighter group to a heavier one). In order to ensure fairer assessments, Japan plans to fortify future standards with the CAFE approach for 15 gross vehicle weight groups. Fourth, the penalties for non-implementation are relatively loose; the energy-saving law is not intended to regulate manufacturers, but to promote their efforts to improve the energy efficiency of each product.

In the Philippines, the Department of Energy promoted the Energy Efficiency Conservation Plan and Program (2005–14) (Tamang 2011). Fuel efficiency information has been voluntarily provided to consumers by automobile manufacturers since 2002. The fuel economy run measures real fuel consumption on a selected road. The results are disseminated online by the Department of Energy. Notably, the results have been known to differ from the corresponding values in the vehicles' catalogs (DOE Phillipines 2006).

8.2.5. New energy vehicles

Like Japan, the eight developing countries have introduced electric vehicles (EVs). Hybrid EVs are available in China, Malaysia, South Africa, Thailand, and Japan. The prices of these new energy vehicles are generally subsidized, and they are exempt from many taxes in order to promote their market penetration. However, new energy vehicles share some common issues concerning market penetration. This section describes the Japanese and Chinese cases as examples.

FCEVs will be on sale in Japan beginning in 2015. The Japanese government plans to have 100 hydrogen fueling stations in 2015 in four major cities and along highways. The government plans to introduce 2 million FCEVs and 1,000 hydrogen fueling stations in 2025 in 47 prefectural capitals. However, the high costs of construction for such fueling station infrastructure may pose problems.

FCEVs are currently being tested in China on a project basis. In 2008, China's Department of Science and Technology implemented a dissemination policy known as the "Ten Cities, Thousand Vehicles" program for plug-in hybrid vehicles (PHEVs) and EVs. Under this program, the Beijing government offered subsidies worth Chinese currency renminbi (RMB) 100,000 for each PHEV and RMB 120,000 for each EV. The Shanghai government provided subsidies of RMB

80,000 for each PHEV and RMB 100,000 for each EV (CATARC 2013). Another new policy for hybrid EV, EV, and FCEV dissemination also provides financial support for purchasing these vehicles. These subsidies can cover up to 60 percent of a vehicle's price. This particular policy is implemented by the Department of Finance, National Taxation Bureau, and the Department of Industry and Communication. In 2011, the combined share of EVs, hybrid EVs, and FCEVs was 0.04 percent of all new car sales, lower than the government's expectations of 2.3 to 2.5 percent. These numbers indicate that the dissemination of new energy vehicles is still in the early stages, and the government needs to provide not only financial but also infrastructure support, such as EV fueling stations.

8.2.6. Vehicle inspection

An effective inspection system is needed to phase out old/inefficient cars. All eight countries have legislation pertaining to vehicle inspection systems. A well-functioning inspection system would ensure that used vehicles are regularly checked for defective vehicle parts, which would then be replaced. Among other things, it would also ensure that the vehicles' engines are tuned. Table 8.6 shows intervals for vehicle inspection by vehicle type for the eight countries. Inspection of commercial vehicles is given top priority, whereas that of passenger cars is not implemented as intensively. Private vehicles must be inspected after they reach a certain age in some countries (Hirota 2010d). Indonesia, Malaysia, and South Africa do not require passenger cars to be inspected.

Brazil has annual inspections for all kinds of vehicles. In China, EURO 3–equivalent passenger cars produced in 2008 or later are required to be inspected once every two years, beginning three years after their first registration. Passenger cars older than 15 years must pass inspection every year. The inspection interval is shorter for EURO 1 and EURO 2 vehicles than for EURO 3 vehicles. In India, inspection is required when the vehicle completes 1 year and 15 years of service. Passenger cars need to re-register after they complete 15 years. Once a re-registered vehicle passes inspection, the registration is valid for another five years. In India, a supplementary roadside inspection called "Pollution Under Control," or PUC, is also mandated. Commercial vehicles must pass this inspection. There are no such mandatory rules for passenger cars in Indonesia and Malaysia. In the Philippines, passenger cars are to be inspected every year. However, as there is no inspection center, only a document of inspection is submitted to renew a vehicle's registration. In Thailand, passenger cars with seven seats must pass inspection every year once the vehicle completes seven years.

A better inspection system would definitely phase out high-emission vehicles and facilitate their replacement with eco-cars. According to Society for International Cooperation, the Government of Germany (GTZ, formerly GIZ, 2005), vehicles that did not pass an inspection test emitted 1.7 to 7 times higher CO levels than vehicles that did. As a successful example, in 2000 a voluntary inspection and maintenance bus project in Jakarta called the "Blue Sky Project," reduced hydrocarbon, CO, and soot emissions by 49 percent, 53 percent, and 61 percent, respectively, and increased fuel savings by 5 percent.

Table 8.6 Inspection interval for passenger cars

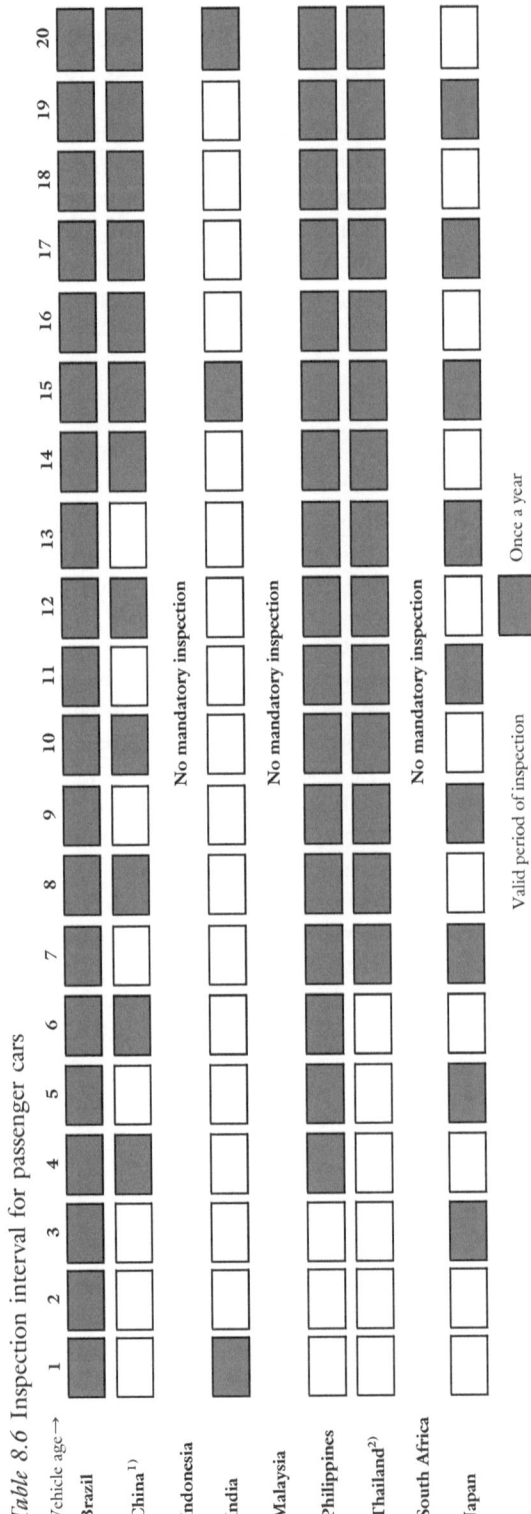

Data sources: ACEA (2001), ADB/ASEAN (2004), DLT Thailand (2001), Ganguli (2010), Ishak, A. (2001), JASIC (2002), LTO Philippines (2006), LTO Thailand (2002), MOI Indonesia (2005), Pham (2001), Phimphun (2002), Sager (2001), Stechdaub (2001), UNESCAP-JARI (2001), and Yuan (2005).

Prepared by the author.

8.3. Do eco-car incentives influence the passenger car market? Evidence of five popular models from each country

8.3.1. Registrations and sales

Many emerging economies have introduced sound environmental sustainability policies for the automobile sector. These policies promote green products, making them more competitive in the marketplace. An analysis of the number of vehicle registrations and new passenger car sales provides useful information. The increase in vehicle registrations is one of the reasons for increased fuel consumption. Figure 8.1 shows the number of passenger cars per 1,000 people from 2005–10 for each country. The Philippines has the lowest rate of passenger cars per 1,000 people. Policy implementation in large countries, such as China and India, continues to exert relatively significant impacts on emission and fuel economy regulations. Figure 8.2 shows details regarding vehicle registration by vehicle type for each country. Passenger vehicles account for more than 50 percent of all vehicles except in the Philippines and Thailand. However, these two countries have recorded higher sales of new passenger cars compared to the others (Figure 8.3), and thus are likely to see an increase in fuel consumption in the future. Consequently, they would need to exercise stringent emission controls and introduce better fuel economy standards.

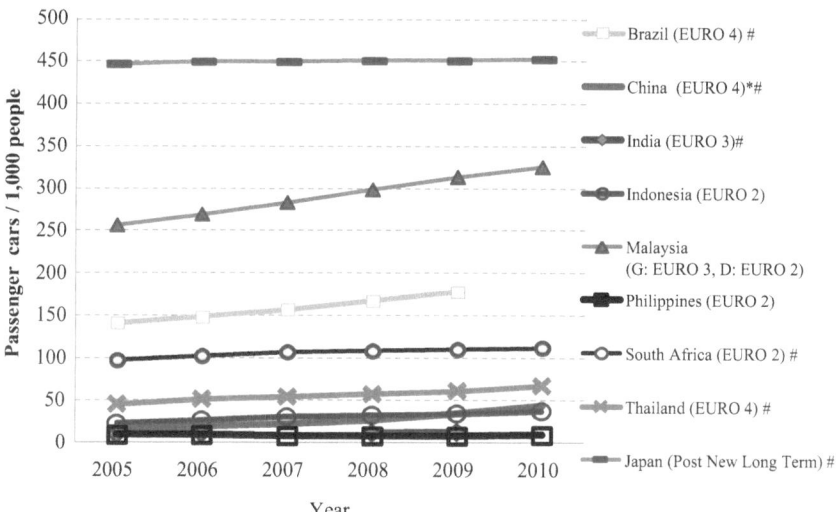

Figure 8.1 Passenger cars per 1,000 people 2005–10

* Hong Kong, Macao, and Chinese Taipei are not include in the Chinese data.

Fuel economy regulation has been introduced/will be introduced.

Data sources: IRF (2012), World Bank (2014).

Prepared by the author.

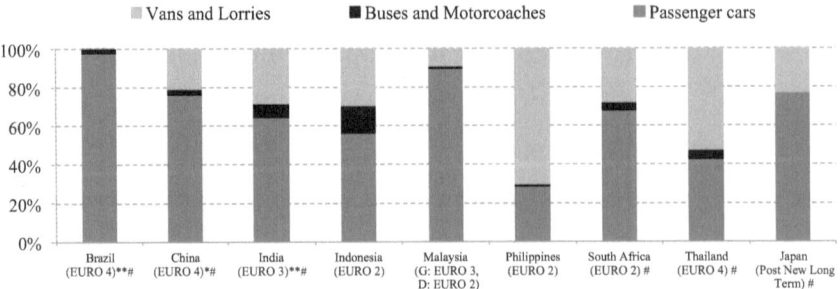

Figure 8.2 Vehicle registration by vehicle type and by country 2009–10

* Hong Kong, Macao, and Chinese Taipei are not include in the Chinese data.

** 2009 data.

Fuel economy regulation has been introduced/will be introduced.

Data sources: IRF (2012), World Bank (2014).

Prepared by the author.

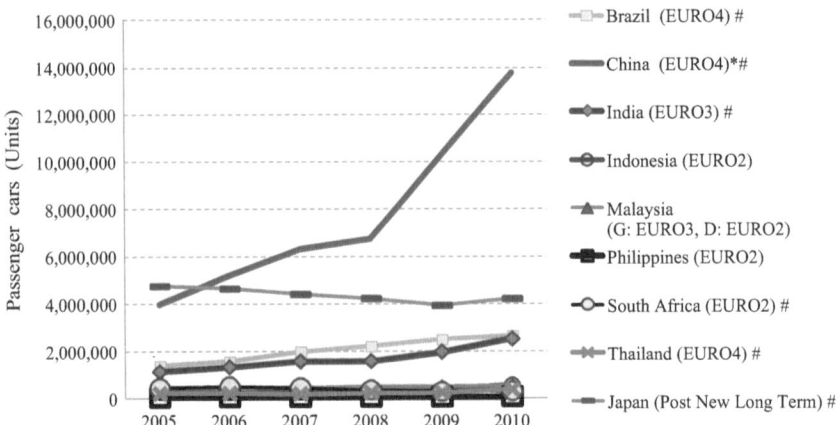

Figure 8.3 Sales of new passenger cars 2005–10

( ): Current emission standard for passenger cars in 2014.

* Hong Kong, Macao, and Chinese Taipei are not include in the Chinese data.

Fuel economy regulation has been introduced/will be introduced.

Data sources: AAF (2014), Ahmad (2010), GAIKINDO (2014), NAAMSA (2014), Nacua and Mills (2010b), and SIAM (2014).

Prepared by the author.

8.3.2. Structure of vehicle tax systems and incentives for eco-cars

The total cost of an eco-car consists of the price of the car, taxes, and compulsory and voluntary insurance. Table 8.7 shows the tax incentives and disincentives for eco-cars at the acquisition stage. The taxes are categorized into three

Table 8.7 Eco-car incentive and disincentive in tax structures at the acquisition stage

Legend: ▢ Eco car incentive ▮ Disincentive

Costs / Country	Brazil	China	India	Indonesia	Malaysia	Philippines	South Africa	Thailand	Japan
Insurance	Voluntary / Compulsory	Voluntary / Compulsory	Voluntary / Compulsory	Compulsory	Compulsory + voluntary	Voluntary / Compulsory	Voluntary	Voluntary / Compulsory	Voluntary / Compulsory
Vehicle acquisition tax	Ownership tax	Acquisition tax; Vehicle consumption tax; Vehicle tax; **Number plate**; **Subsidies EV**	Registration	Luxury tax 0% LCGC	Registration; Loan commission; Number plate; Road tax; Inspection fee	Acquisition	CO_2 Tax 90 Rand each g/km when catalogue CO_2 emission exceeds 120 g/km.	Registration	Acquisition 50%–100% tax reduction EV,PCHV, CNG, exceed FE 2015 level and lower emission than 2005 standard
Local tax	ICMS								
VAT	PIS /COFINS	VAT	VAT	VAT	Sales tax	VAT	VAT	VAT	Consumption tax
Excise duty			Excise duty			Excise duty		Excise duty 30% → 17% CO_2 ≤ 100 g/km , Fuel consumption ≤ 4.3 l/100km EURO4 Engine capacity Gasoline cars ≤1300 cc, Diesel cars ≤ 1500 cc	
Central gov't tax								Interior tax	
Import tariff	Import tariff; IPI 13% →11% FPV 1000cc-1999cc; IPI INOVAR-AUTO 30 % tax reduction	Import tariff	Import tariff HEV component 10%→6%	Import tariff	Import tariff HEV : 0 %	Import tariff EV,HEV,CNG components 0%; Import commission fee	Import tariff; Ad valorem	Import tariff	Import tariff

Data sources: ACFA (2012, 2014), Bathan (2010), *China Daily* (2012), Ding (2010), Hirota (2004, 2006, 2007a, 2007b, 2008a, 2008c, 2008d, 2010a, 2010b, 2010c), Infante (2011), JETRO (2012), Kovongpanich (2014), MAE (2014), MEP (2013), Mills and Nacua (2011), MNRE (1996a, 1996b), Nacua and Mills (2010b), Rayner (2010), Yasin and Zainudin (2011), and Zakaria (2013).

Prepared by the author.

types: import/central government tax/excise duty, consumption tax, and vehicle-related tax. In general, the import/central government tax/excise duty and consumption taxes are included in the catalog price. Vehicle-related taxes and costs are paid by the vehicle user. Indonesia and Malaysia are exceptions.

Table 8.7 illustrates considerable variations in the acquisition taxes and other costs imposed on private vehicles across the eight countries. The levels of import tariff/central government tax/excise duty, value-added tax (VAT), and local tax are imposed on the basis of the asset's economic value. When taxing cars based on their market value, the incentive structure does not respond to environmental concerns. Brazil, India, Malaysia, the Philippines, and Thailand offer tax reductions or eliminate taxes for specific vehicle categories (fuel type, new energy vehicle, engine capacity) on import tariffs or excise duties. Brazil offers a tax incentive via a 2 percent tax reduction on Fuel Flex Vehicles (FFV) with engine capacities ranging from 1000cc to 1999cc and a 30 percent tax reduction to car manufacturers that fulfill the conditions of domestic production, investments in R&D and production technology, and fuel economy improvement. India reduced its import tariff on hybrid electric vehicle (HEV) components. Malaysia provides tax relief for promoting EVs. The Philippines offers import tariff relief for EVs, HEVs, and compressed natural gas (CNG). Thailand established a two-phase eco-car program, with Phase 1 starting in 2007 and Phase 2 in 2013. Thailand has reduced its excise duty from 30 to 17 percent. There is no tax incentive on VAT nor a local tax among the eight countries.

A vehicle acquisition tax based on engine capacity/vehicle weight could be a potential incentive to purchase or maintain vehicles with smaller engine capacities/lighter weights, provided the governments impose lower tax levels for more efficient/less polluting vehicles. China (Shanghai) provides a disincentive by imposing a high charge in terms of a number plate fee for private vehicles. China also provides subsidies on EVs. Indonesia eliminates the luxury tax for vehicles priced below 95 million Rupiahs and for vehicles with engines that use high-quality fuels. South Africa introduced an environmental levy on new passenger cars with CO_2 emissions exceeding 120 g/km.

8.3.3. Results by country

This section calculates the competitiveness of eco-cars by estimating their total costs at the acquisition stage. A large city was selected from each of the large countries included in this study: São Paulo (Brazil), Shanghai (China), Bangalore (India), and Jakarta (Indonesia). Five passenger car models were selected from each of the abovementioned countries depending on the top-selling cars in that country's market. While Models D and E are the same for the eight countries and Japan, their engine capacities are different due to limited options from among the available car models. If there was no description of fuel type, it was assumed to be a gasoline vehicle. For models with options for engine capacities, we selected different engine capacities, because we wanted to estimate variations in the tax level by engine capacity. All the taxes and costs were calculated

Table 8.8 Selected models by country, eco-car incentive and disincentive

			Eco car incentive
			Disincentive
		● ○	Domestic production

Brazil	●Model A 1000cc	●Model B 1400cc	Model C 1600cc	●Model D 2000cc	●Model E 1800cc
China (Shanghai)	●Model F 1100cc	●Model G 2000cc	Model H 3000cc	●Model D 1800cc	●Model E 1800cc
India	●Model L 800cc	●Model M 2500cc	Model N 2000cc	●Model D 1800cc	●Model E 1800cc
Indonesia	●Model I 1500cc	●Model J 1000cc	Model K 1200cc	Model D 2400cc	●Model E 1800cc
Malaysia	●Model O 1300cc	●Model P 1500cc	Model Q 1500cc	Model D 1600cc	●Model E 1800cc
Philippines	●Model F 1500cc	●Model S 3200cc	Model T 1400cc	Model D 1800cc	●Model E 1600cc
South Africa	●Model U 1400cc	●Model V 3000cc	Model X 1500cc	Model D 1800cc	●Model E 2000cc
Thailand	●Model X 2500cc	●Model Y 1500cc	Model Z 2000cc	●Model D 1800cc	●Model E 1800cc

Prepared by the author.

according to regulations prevalent in 2013. Vehicle prices, taxes, and costs were revised and verified by our local vehicle experts, engineers, and local authorities. The comparative analysis allows an overall assessment of each eco-car in terms of its cost at acquisition.

Table 8.8 shows selected models by country and incentives and disincentives of owning an eco-car. If a model specification meets the eco-car requirement, we apply the tax reduction to the calculation. A disincentive is indicated by the implementation/presence of a tax policy detrimental to environmental improvement.

(1) Brazil

Inovar-Auto is a tax-incentive scheme in Brazil. Inovar-Auto offers a 30 percent tax reduction to car manufacturers that fulfill the conditions of domestic production, investments in R&D and production technology, and fuel economy improvement. The program, however, could help develop the automobile industry in Brazil. There are no specifications unique to different models; thus, all the models fulfilling the conditions can avail of the 30 percent tax reduction under Inovar-Auto (MAE 2014) (Figure 8.4). It is likely that car manufacturers will become very competitive if Inovar-Auto were to be implemented, because the program bears the biggest share of the total cost of a vehicle at the

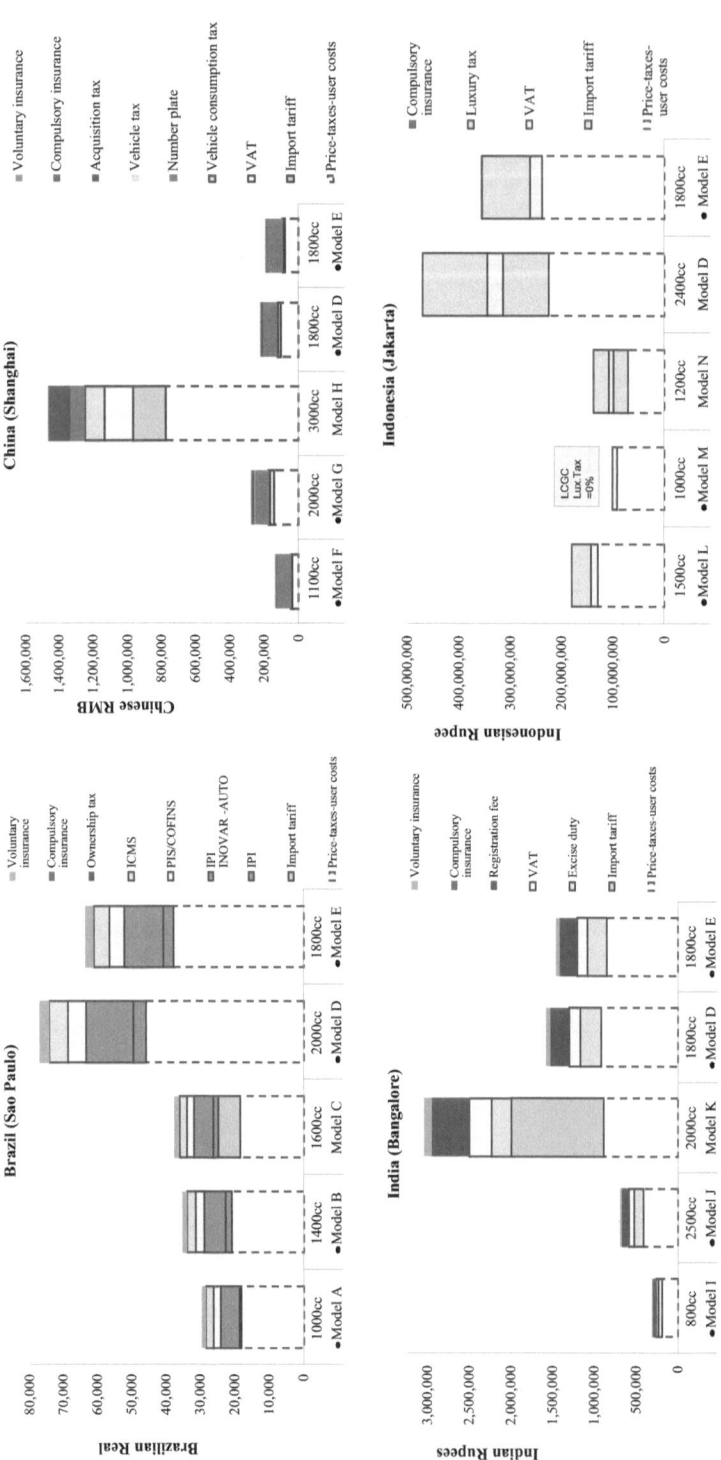

Figure 8.4 Total costs at acquisition: Brazil, China, India, and Indonesia

Data sources: Central Government of China (2000), Chandrasa (2013), CTAX (2014), Fedex World Tariff (2013), iCarros-Revista (2013), IRDA (2012), Labhsetwar (2013), Marathe (2013), Nozaki (2012), Orikassa (2013), Secretaria da Fazenda, Governo de Sao Paulo (2014), Shanghai Government (2014), Shanghai International Commodity Auction Co., Ltd. (2012), SIAM (2013), Sonpo Soken (2013), Suweden (2013), Transport Department Government of Kartanaka (2010), Xu (2013).

Prepared by the author.

acquisition stage. The results show that Model D is the costliest model at the acquisition stage in Brazil.

(2) Shanghai

The government provides a disincentive by imposing a high charge in terms of a number plate fee for private vehicle registrations. The fee for a number plate is determined by the number of applications received, and thus changes every month. The fee is the same for all models. Model H is the costliest at the acquisition stage, as it has the most powerful engine and is the most expensive of the five models (Figure 8.4).

(3) India

The Bangalore government sets the registration tax according to the market value of the vehicle. This may serve as a disincentive for car replacements/ phase-outs. Most models are locally produced even if they are foreign brands so as to avoid a high import tax (Figure 8.4).

(4) Indonesia

Indonesia's Low Cost Green Car (LCGC) policy provides a luxury tax relief equal to 10 percent of the car's price for certified green cars. The price of the car must not exceed Indonesian 95 million Rupiahs. The engine must use high-quality gasoline or diesel. Model M is certified as an LCGC. It is the most competitive among the five models (Figure 8.4). Its insurance fee is quite low, but not all vehicles are covered by insurance in Indonesia.

(5) Malaysia

The Malaysia Energy Commission ("Pusat Tenaga Malaysia") recognizes the importance of tax relief in its roadmap for promoting EV use. The tax relief is envisaged to target infrastructure industries, automobile manufacturers, and users. Various tax incentives are provided at the acquisition stage. For example, import duties on EVs and hybrid EVs were waived until the end of 2013. Model Q is tax-free at import, but its total cost is higher compared to that of Model P due to its high car price even after excluding the tax (Figure 8.5). Malaysia also provides a subsidy of Malaysian Ringgit 5,000 (approximately $1,652) for a new vehicle – a Proton or a Perodua – when upgrading vehicles that are 10 years or older and for which road taxes and loans have been paid.

(6) The Philippines

There are no incentives for eco-cars in the Philippines. Model T and Model D are competitively priced even though they are both imported (Figure 8.5).

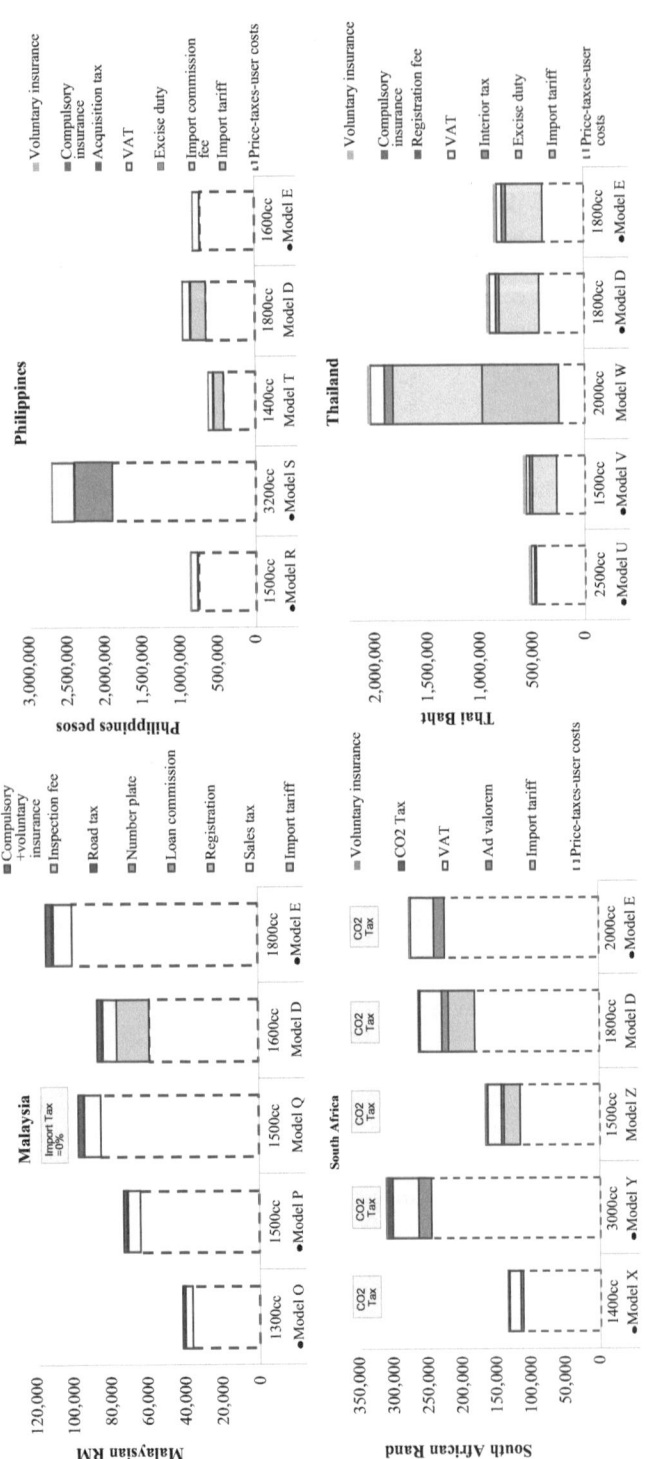

Figure 8.5 Total costs at acquisition: Malaysia, Philippines, South Africa, and Thailand

Data sources: Addrus (2013), BOI Thailand (2013), Custom Department Thailand (2013), Fedex World Tariff (2013), Government of the Philippines (2000), MAI (2012), Matrade (2014), Mbandi (2013), Olaguer (2013), Sonpo Soken (2013), SARS (2009), SARS (2010), SARS (2013), Suwanvithaya (2013), Techawiboonwong (2014).

Prepared by the author.

(7) South Africa

In September 2010, South Africa introduced an environmental levy on new motor passenger cars with CO_2 emissions exceeding 120 g/km. There is no compulsory insurance tax in South Africa (Figure 8.5).

(8) Thailand

The Board of Investment (BOI) under the Ministry of Industry established the project "Eco-car" in 2007. Phase 1 (2007–10) was intended to promote invest-ments for car manufacturers focusing on emission limits and fuel consumption apart from safety concerns (i.e. front and side crash tests). In 2013, the BOI announced Phase 2 of the project, wherein interested manufacturers would need to meet more stringent requirements for eco-cars, in terms of both emission levels and fuel consumption, before 31 March 2014. However, none of the five models can be classified as eco-cars, because the criteria for eco-cars places limits on engine capacity. Besides reducing the excise duty on eco-cars from 30 to 17 percent, the Thai government offers other import tax incentives for pickup trucks, and import taxes on single cabs have been reduced to 3 percent. Most of the total cost of Model W, an imported car, appears to consist of tax payments (Figure 8.5).

8.3.4. Results for the common models

Figures 8.6 and 8.7 show the total costs at the acquisition stage for the most common models – Model D and Model E, respectively – for all eight countries and Japan. As the comparison is along international lines, the total costs are understandably diverse. The different total costs reflect the differences in the vehicular systems of the countries. Domestic production enables manufacturers to avoid import tariffs. The results indicate limited eco-car uptake in the eight developing countries. Model D qualifies as an eco-car only in South Africa. Model E qualifies as an eco-car in South Africa and Japan. The total acquisition taxes reveal higher vehicle registration taxes than in Japan. In Brazil and Thailand, manufacturing/import taxes are a higher share of the total costs at the acquisi-tion stage for Model D and Model E. In Shanghai, the vehicle registration fee is a high share of the total cost at acquisition because it uses high vehicle reg-istration taxes as a disincentive against car ownership. Indonesia has larger share of registration tax in total costs because of its luxury tax. The registration taxes imposed by India are detrimental to environmental protection because they depend on a vehicle's age.

8.4. Conclusion: what are the elements of a successful eco-car policy?

This chapter compared the eco-car policies of new passenger cars adopted by eight developing countries and Japan. The results show that the eight countries face different challenges with regard to promoting eco-cars (new cars) in their

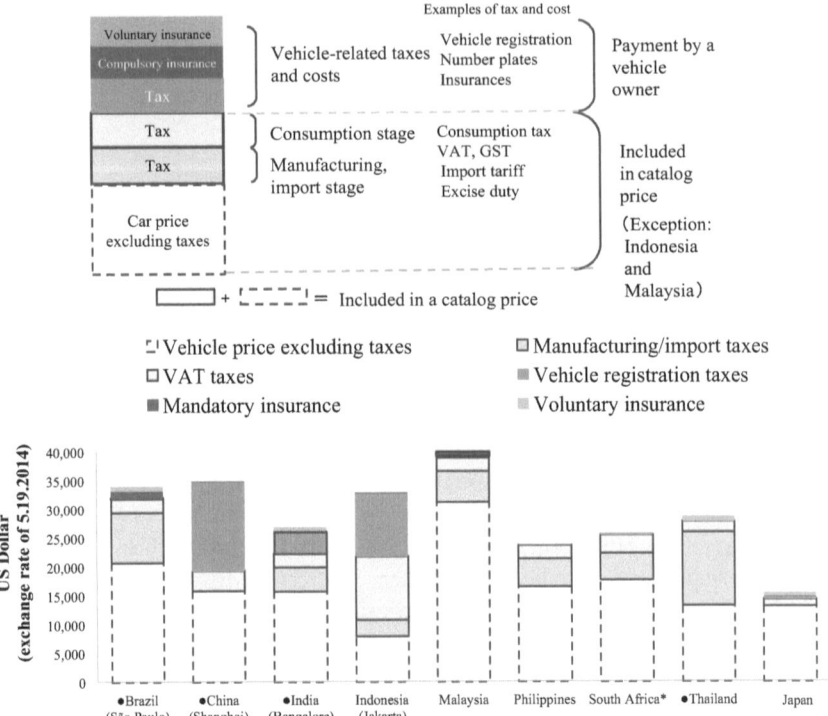

Figure 8.6 Total costs at the acquisition stage: Model D

• Domestic production

* Eco-car

Data sources: Data sources: Addrus (2013), BOI Thailand (2013),Custom Department Thailand (2013), Fedex World Tariff (2013), Government of the Philippines (2000), MAI (2012), Matrade (2014), Mbandi (2013), Olaguer (2013), Sonpo Soken (2013), SARS (2009), SARS (2010), SARS (2013), Suwanvithaya (2013), Techawiboonwong (2014).

Prepared by the author.

respective countries. It is hoped that Japan's experience in eco-car policy implementation will serve as a guidance for the developing countries hoping to expand eco-car usage. Finally, this chapter attempted to answer to the question, "what are the elements of a successful eco-car policy?" In summary, it is important to reform the entire vehicle-related system and its associated infrastructure to help disseminate eco-cars. Eco-car policies should ideally provide a wide range of criteria governed by simple indicators. Japan's experience in this area provides three important and realistic suggestions.

First, the technical measures reviewed in the previous sections are interconnected; each measure should be implemented in tandem with the remaining technical measures. Section 8.2 compared the technical abilities (i.e. registration, emission standard, fuel economy, xEVs, and mandatory inspection) of the eight countries and Japan. The first element of policy implementation should concern

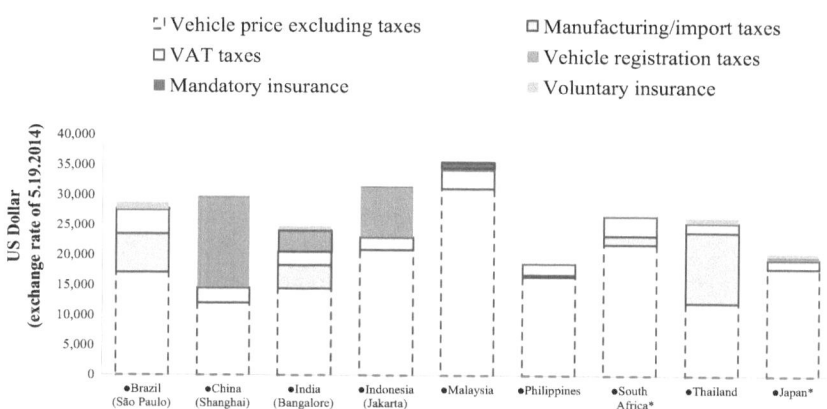

Figure 8.7 Total costs at the acquisition stage: Model E

• Domestic production

* Eco-car

Data sources: Data sources: Addrus (2013), BOI Thailand (2013),Custom Department Thailand (2013), Fedex World Tariff (2013), Government of the Philippines (2000), MAI (2012), Matrade (2014), Mbandi (2013), Olaguer (2013), Sonpo Soken (2013), SARS (2009), SARS (2010), SARS (2013), Suwanvithaya (2013), Techawiboonwong (2014).

Prepared by the author.

vehicle registration. The availability of more detailed data would allow for a more quantitative analysis for accurate assessment and estimation. Data for the number of vehicles owned by vehicle type (passenger cars, commercial vehicles, and motorcycles) are available for all eight countries, including Japan. However, data for registration by vehicle age, fuel type, renewal, and deletion are available for Japan and certain countries only. Among the eight countries, Japan was the first to introduce stringent emission standards. Distribution of good quality fuel is a common issue faced by developing countries with regard to emission standards. It is prudent for large countries to introduce more stringent emission standards at the city level first. The time schedule for implementing the emission standards needs to be revised from the viewpoint of fuel refinery upgrades. Beijing's incentive for replacing old cars by implementing a labeling system seems to be very effective. The comparative study of fuel economy regulations showed a great deal of diversity in the design of the standards. Four of the key features of Japan's fuel economy regulation are noteworthy: the Top Runner approach, periodic update of the standard, categorization to avoid unfair assessments, and loose penalties for car manufacturers. Japan and the eight countries have introduced EVs. Hybrid EVs are available in four of these countries and Japan. FCEVs are currently being tested in China on a project basis and will be on sale in Japan beginning in 2015. These new energy vehicles are generally subsidized and exempt from many taxes, in line with the eco-car policies. However, the dissemination of new energy vehicles seems to be in the early stages. The government needs to provide not only financial support but also infrastructure (e.g. a well-functioning inspection system) to increase their uptake. Inspection of commercial vehicles is given top priority, whereas that of passenger cars is not implemented as intensively, and this practice needs to change. Phasing out older cars and replacing them with newer/more efficient vehicles is very important.

Second, it is crucial that eco-car incentives be designed and implemented appropriately. Section 8.3 showed the eco-car incentive and disincentive in tax structures at the acquisition stage. Eco-car incentives include tax reductions, tax-free programs, and subsidies. Tax disincentives control the number of cars owned. The levels of import tariff/central government tax/excise duty, value-added tax (VAT), and local tax are imposed on the basis of the asset's economic value. When taxing cars based on their market value, the incentive structure does not respond to environmental concerns. Brazil, India, Malaysia, the Philippines, and Thailand offer tax reductions or elimination of taxes for specific vehicle categories (FFV, HEV, EV, CNG and their components, smaller engine capacity) on import tariffs or excise duties. A vehicle acquisition tax based on engine capacity/vehicle weight could be a potential incentive to purchase or maintain vehicles with smaller engine capacities/lighter weights, provided the governments impose lower tax levels for more efficient/less polluting vehicles. China (Shanghai) provides a disincentive for private vehicle ownership and subsidizes EVs. Indonesia eliminates the luxury tax for certified vehicles with low prices and engines that use high-quality fuels. Section 8.3 compared the competitiveness of eco-cars by estimating the total costs of five models at the acquisition stage. The results indicated that although these models are top sellers in their respective countries, the majority are not certified as eco-cars. An ideal eco-car policy, therefore, should meet a wide range of criteria governed by simple indicators. For example, South Africa applies a vehicle registration fee depending on the vehicle's CO_2 emission rate, not per the model. Japan applies an indicator that integrates the emission standard and the fuel economy standard. The other countries classify eco-cars depending on limited, but specific, technical and/or price requirements.

Third, all the developing countries reviewed in this chapter share similar characteristics, such as weakness in government oversight and inadequate infrastructure and implementation, which stem from inadequate human resources and facilities for monitoring environmental performance of vehicles. Shortfalls in resource allocation and coordination may be improved in two ways: revenue allocation and information sharing.

The revenue earned from availing eco-car incentives should be allocated toward developing human resources and upgrading measurement facilities. Following Japan's example, the countries would benefit greatly from information sharing, setting quantitative targets, and improving cooperation among different stakeholders; in fact, these elements helped Japan achieve its targets earlier than planned. Thus, coordination and agreement among stakeholders will be a key issue for developing countries hoping to implement successful eco-car policies in their respective countries.

Acknowledgments

This chapter was conducted as a part of the research for an acquisition tax estimation project. I am grateful to my colleagues Ms. Mami Igarashi and Ms. Haruka Ohtani for advice on data collection and calculation. I appreciate Prof. Shigeru Matsumoto of Aoyama Gakuin University and many colleagues, including Dr. Shogo Sakamoto, Dr. Masaaki Fuse, Dr. Hisashi Imanaga, Dr. Yoichi Shimakawa, and Dr. Hiroyuki Kosaka from Kashima Seminar of Chuo University, for discussion. We are solely responsible for all the errors in this chapter.

References

Addrus, A. (2013) "Costs of car ownership in Malaysia." Email (Personal communication, 11–12 December 2013).

Ahmad, A.D. (2010) "Sales statistics." Email and interview (Personal communication, October–December 2010).

Andrianto, J. (2008) "Emission regulation in AAF countries." Email (Personal communication, 30 January 2008).

ASEAN Automotive Federation (AAF) (2008) "Emission and fuel schedule." Email (Personal communication, 30 January 2008).

ASEAN Automotive Federation (AAF) (2014) *Statistics.* Online. Available: <http://www.asean-autofed.com/> (Accessed 17 August 2014).

Asian Clean Fuels Association (ACFA) (2012) *Emission Regulation in Beijing.* Online. Available: <http://www.acfa.org.sg/newsletterfuelforthought11.php?question=2#question2a> (Accessed 18 May 2012).

Asian Clean Fuels Association (ACFA) (2014) *Emission Regulation in Malaysia.* Online. Available: <http://www.acfa.org.sg/newsletterinfocus08_01.php> (Accessed 17 August 2014).

Associação Nacional dos Fabricantes de Veículos Automotores (ANFAVEA) (2014) Estatística (Statistics) (in Portuguese) Online. Available: <http://www.anfavea.com.br/tabelas.html> (Accessed 25 March 2015).

Association des constructeurs européens d'automobiles (ACEA) (2001) *Motor vehicle taxation, 2001 Edition.* Brussel: ACEA.

Asian Development Bank (ADB)/Association of Southeast Asian Nations (ASEAN) (2004) *Malaysia Road Safety Action Plan, ADB/ASEAN Regional Road Safety Program.* 32. Online. Available: <http://jpj.gov.my/bisnesf.htm, http://www.jpj.gov.my/com_rtax.htm> (accessed 8 February 2001).

Association of Indonesia Automotive Industries (GAIKINDO) (2014) *Statistical Data.* Online. Available: <http://gaikindo.or.id/index.php?option=com_content&task=blogcategory&id=0&Itemid=145> (Accessed 16 August 2014).

Bathan, G. (2010) *The Philippines Issues EURO4 Vehicle Emission Standard.* Online. Available: <http://cleanairinitiative.org/portal/node/6362> (Accessed 17 August 2014).

Board of Investment (BOI) Thailand (2013) *Import Tariff in Thailand.* Online. Available: <http://www.boi.go.th/index.php?page=opp_automotive&language=en> (Accessed 15 November 2013).

Central Government of China (CGC) (2000) *Vehicle Tax.* Beijing: State Council. Online. Available: <http://www.gov.cn/gongbao/content/2001/content_60852.htm > (Accessed 4 February 2014).

Chandrasa, G.T. (2013) "Tax in Indonesia." Email (Personal communication, 21 October 2013).

Chen, C. (2009) "Automotive emission control and improvement of air quality in Shanghai," in *JARI China Roundtable 2009*, Beijing, 13 February. Tsukuba: JARI.

China Automobile Technology and Research Center (CATARC) (2013) *China Auto Industry Yearbook*. Tienjin: CATARC.

China Daily (2012) "Emission Regulation in Beijing." Online. Available: <http://www.chinadaily.com.cn/cndy/2012–05/18/content_15326070.htm> (Accessed 18 May 2012).

Clean Air Initiative-Asia (CAI-Asia) (2009) *Emission Standard by Country*. Online. Available: <http://cleanairinitiative.org/portal/SynovateEvalReport> (accessed 5 March 2009).

Cooperación Técnica Alemana (GTZ, formerly GIZ) (2005) El Impacto Ambiental de la Revisión Técnica Vehicular:Su aporte en el control de emisiones vehiculares contaminantes. Estudio realizado para el Gran Área Metropolitana Periodo 2003-2005 (Vehicular Technical Review: Its contribution to pollution control vehicle emissions. Study for the Greater Metropolitan Area Period 2003-2005) (in Spanish). Online. Available HTTP: http://www.nacion.com/ln_ee/2007/enero/06/gasolina.pdf (Accessed 8 June 2007).

CTAX (2014) *Vehicle Consumption Tax*. Online. Available: <http://www.ctax.org.cn/news/rdzt/qcxfstz/> (Accessed 12 February 2014).

Custom Department (CD) Thailand (2013) *Tax Calculation*. Online. Available: <http://www.customs.go.th> (accessed 15 November 2013).

Delphi (2014/2015) *Worldwide Emission Standards, Delphi*. Online. Available: <www.delphi.com/emissions-pc> (Accessed 25 April 2014).

Department of Energy (DOE) of the Philippines (2006) *Fuel Economy Runs*. Online. Available: <http://www.doe.gov.ph/EE/Fuel%20Economy%20Run.htm> (Accessed 9 November 2011).

Department of Land Transport (DLT) of Thailand (2001) "Motor vehicle inspection and taxation in Thailand." Email (Personal communication, 10 January 2001).

Ding, Y. (2009) "Emission reduction regulation in China," in *JARI China Roundtable 2009*, Beijing, 13 February. Tsukuba: JARI.

Ding, Y. (2010) "Vehicle emission regulations in China." Interview (31 August 2010).

Fedex World Tariff (FWT) (2013) *Database*. Online. Available: <https://ftn.fedex.com/wtonline/jsp/wtoMainUL.jsp?> (Accessed 15 October 2013–14, February 2014).

Ganguli, A. (2010) "Low carbon development strategy in road transport sector," in *JARI India Seminar*, Delhi, 10 March. Tsukuba: JARI.

Gao, H. and Jin, Y. (2011) "Progress of China auto fuel economy standard," in *JARI China Roundtable*, Beijing, 2 November. Tsukuba: JARI. p. 5.

Global Fuel Economy Initiative (GFEI) (2014) *Fuel Economy State of the World 2014*. Paris: GFEI.

Government of the Philippines (GP) (2000) *Republic Act 8794*: An act imposing a motor vehicle user's charge on owners of all type of motor vehicles and for other purposes. 27 June 2000.

Hirota, K. (2008a) "Report of JARI China roundtable 2008 – Environmental issues and transportation policies at city level," *JARI Research Journal*, 30: 39–42.

Hirota, K. (2008b) "Report on the 10th Meeting of AMEICC WG-AI-AEM-METI economic and industrial cooperation committee working group on automobile industry," *JARI Research Journal*, 30: 21–6.

Hirota, K. (2008c) "Report of JARI Shanghai fuel quality roundtable 2008," *JARI Research Journal*, 31: 33–8.

Hirota, K. (2008d) "Report of JARI Indonesia roundtable 2008 – Efforts towards air pollution reduction in Indonesia," *JARI Research Journal*, 30: 39–44.

Hirota, K. (2008e) "Report on the 10th Meeting of AEM-METI economic and industrial cooperation committee working group on automobile industry (AME-ICC WG-AI)," *JARI Research Journal*, 30: 63–4.

Hirota, K. (2009) "Policy for better air quality in Asia: proposal for a policy evaluation method for four ASEAN countries," *Studies of Regional Science*, 38: 1093–104.

Hirota, K. (2010d) "Comparative studies on vehicle-related policies for air pollution reduction in Asian 10 countries," *Sustainability*, 2: 145–62. Online. Available: <www.mdpi.com/journal/sustainability> (Accessed 8 January 2010).

Hirota, K. (2010a) "Report of JARI China Roundtable 2009 – Vehicle emission and fuel economy regulations in the future," *JARI Research Journal*, 32: 35–42.

Hirota, K. (2010b) "Fuel economy regulation in China and efforts for emission reduction in Beijing – JARI China roundtable 2010," *JARI Research Journal*, 33: 29–33.

Hirota, K. (2010c) "Report of Indo-JARI roundtable 2009 – Fuel quality and vehicle emission in India," *JARI Research Journal*, 32: 43–51.

Hirota, K. (2012) "Recommendations regarding the status of data collection concerning automobiles in East Asian countries," in JARI (ed.) *Survey Analysis of the Road Transport Sector for Reducing CO_2 Emissions, ERIA Report*. Tokyo: The Commission of the Ministry of Economy, Trade and Industry.

iCarros-Revista (iCR) (2013) *Quanto de imposto você paga no carro?* (How much tax do you pay in the car?) (In Portuguese). Online. Available: <http://www.icarros.com.br/noticias/mercado/quanto-de-imposto-voce-paga-no-carro-/14740.html> (Accessed 10 January 2014).

Infante, R. (2011) "Emission regulation in the Philippines." Interview (15 November 2011).

Insurance Regulatory and Development Authority (IRDA) (2012) *Handbook on Motor Insurance*. Online. Available: <http://www.policyholder.gov.in/Motor_Handbook.aspx> (Accessed 17 December 2013).

International Council on Clean Transportation (ICCT) and DieselNet (2014) *Transport Policy Net. Brazil: Light-duty: Emissions*. Online. Available: <http://transportpolicy.net/index.php?title=Brazil:_Light-duty:_Emissions> (Accessed 17 August 2014).

International Road Federation (IRF) (2012) *World Road Statistics*. Geneva: IRF.

Ishak, A. (2001) "Vehicle inspection system in Malaysia," in UN Economic and Social Commission for Asia and the Pacific (UNESCAP) and Japan Automobile Research Institute (JARI) (eds.), *The GITE Regional Workshop on Inspection and Maintenance Policy in Asia*. Bangkok: UNESCAP and JARI.

Japan Automobile Standards Internationalization Center (JASIC) (2002) *Systems Related to Motor Vehicle Safety and Environment in Japan. Country Report 2002*. Tokyo: JASIC.

Japan External Trade Organization (JETRO) (2012) *Emission Regulation in China*. Online. Available: <http://www.jetro.go.jp/mail5/u/l?p=j3-HCN2rkNmLx8bdqlEpWQZ> (Accessed 18 May 2012).

Kovongpanich, W. (2014) "Study of fuel economy standard and testing procedure for motor vehicles in Thailand," in *Automotive Summit*, 20 June. Bangkok: Thai Automotive Institute.

Labhsetwar, N. (2013) "Tax and insurance in India." Email (Personal communication, 16 December 2013).

Land Transport Office (LTO) Philippines (2006) *New Motor Vehicle Registration, Land Transport Office in Philippines: Manila, Philippines*. Online. Available: <http://www.lto.gov.ph/docreq.html> (Accessed 17 February 2007).

Land Transport Office (LTO) Thailand (2002) *Motor Vehicle Inspection and Taxation in Thailand*. Bangkok: LTO.

Malaysian Automotive Institute (MAI) (2012) *Cost of Vehicle Ownership in Malaysia, Indonesia and Thailand*. Kuala Lumpur: MAI.

Marathe, N.V. (2013) "Insurance in India." Email (Personal communication, 13 December 2013).

Matrade (2014) *Malaysian Import Tariff*. Online. Available: <http://www.matrade. gov.my/en/foriegn-buyers-section/70-industry-write-up–services/555-green-technology-services> (Accessed 15 November 2013).

Mbandi, A. (2013) "Tax and insurance in South Africa." Email (Personal communication, 15 November 2014).

Mills, V.T. and Nacua, F. (2011) *Car parc and sales in the Philippines*." Email (Personal communication, 24 November 2011).

Ministry of Environmental Protection (MEP) (2013) "Overview of vehicle emission standards in China," in *Clean Fuels and Vehicles Forum*, 4 November. Singapore: Clean Fuel Vehicle Forum.

Ministry of Industry (MOI) Indonesia (2005) *Directorate General for Transportation Equipment and ICT Industries*. Jakarta: Ministry of Industry of Indonesia.

Ministry of Natural Resources & Environment (MNRE) Malaysia (1996a) *Environmental Quality (Control of Diesel Engines) Regulation*. Putrajaya: DOE, PU(A) 429/1996.

Ministry of Natural Resources & Environment (MNRE) Malaysia (1996b) *Environmental Quality (Control of Emission from Petrol Engines) Regulations*. Putrajaya: DOE, PU(A) 543/1996. Online. Available: <http://cp.doe.gov.my/cpvc/wp-content/uploads/2011/04/Regulations/EnvironmentalQuality(Control_o%20f_Emission_from_Petrol_Engines).pdf> (Accessed 26 March 2015).

Mundo Ambiente Engenheria (MAE) (2014) *Legislasão*. (Legislation) (In Portuguese). Online. Available: <http://mundoambiente.eng.br/new/legislacao/> (Accessed 17 August 2014).

Nacua, F. and Mill, V.T, Jr. (2010a) "Emission regulation in the Philippines." Interview and emails (Personal communication, October–November 2010).

Nacua, F. and Mill, V.T, Jr. (2010b) "Sales statistics." Interview and emails (Personal communication, October–November 2010).

National Association of Automobile Manufactures of South Africa (NAAMSA) (2014) *Industry Vehicle Sales, Production, Export and Import Data 1995–2011*. Online. Available: <http://www.naamsa.co.za/papers/2009_2ndquarter/schedule.htm> (Accessed 7 August 2014).

Nozaki, H. (2012) *Chugoku Jidousya Hoken Shijyou (Insurance Market in China)* (in Japanese). Hoken Mainichi Shinbun Sya (Homai). Tokyo: Homai.

Olaguer, E.A. (2013) "Tax and insurance in the Philippines." Email (Personal communication, 18 December 2013).

Orikassa, E. (2013) "IPI TAX." Email (Personal communication, 10 January 2014).

Pham, Q. T. (2001) "Technical inspection and maintenance system in Vietnam" in UN Economic and Social Commission for Asia and the Pacific (UNESCAP) and Japan Automobile Research Institute (JARI) (eds.), The GITE Regional Workshop on Inspection and Maintenance Policy in Asia. Bangkok: UNESCAP and JARI.

Phimphun, O.A. (2002) *Government Vehicular Agency and a Role to Control Air Pollution.* Bangkok: Engineering and Safety Bureau, Department of Land Transport.

Prawiraatmadja, W. (2008) "Pertamina's fuel supply strategy," in JARI Indonesia Roundtable, Jakarta, 14 February. Tsukuba: JARI.

Presidência da República (PR) (2012) *Decreto* (Decree) (In Portuguese). *Nº 7.819, de 3 de Outubro de 2012.* Online. Available: <http://www.planalto.gov.br/ccivil_03/_ ato2011–2014/2012/Decreto/D7819.htm> (Accessed 10 January 2014).

Rayner, S. (2010) *Fuel Economy/CO$_2$ Labeling and Taxation: South African Motor Industry Experience.* Pretoria: National Association of Automobile Manufacturers of South Africa (NAAMSA)/Ford Motor Company.

Sager, H. (2001) *Clean Air Project Final Report CAP II.* Indonesia: Swisscontact.

Secretaria da Fazenda, Governo de Sao Paulo (SFGSP) (2014) *IPVA (Imposto sobre a Propriedade de Veículo Automotor) (Tax on vehicle ownership)* (in Portuguese). *Anexo I Tabla de valores venais para cálculo do IPVA 2014.* Online. Available: <http://www.fazenda.sp.gov.br/download/ipva/valor_venal_2014.pdf> (Accessed 10 January 2014).

Sengupta, B. (2009) "Former Member Secretary, Air Pollution Reduction Policy, Central Pollution Control Board," in *Indo-Japanese Conference on Fuel Quality and Vehicular Emissions*, New Delhi, 17–18 March.

Shanghai Government (SHG) (2014) *Vehicle Tax in Shanghai.* Online. Available: <http://www.csj.sh.gov.cn/pub/xxgk/zcfg/ccs/201112/t20111229_388931. html> (Accessed 4 February 2014).

Shanghai International Commodity Auction Co., Ltd (2012) *Number Plate Fee.* Online. Available: <http://www.alltobid.com/guopai/contents/56/1972.html> (Accessed 10 December 2014).

Society of Indian Automobile Manufacturers (SIAM) (2013) *VAT.* Online. Available: <http://www.siamindia.com/scripts/siamsuggestionsforvatimplementation.aspx> (Accessed 13 September 2013).

Society of Indian Automobile Manufactures (SIAM) (2014) *Industry Statistics.* Online. Available: <http://118.67.250.203//scripts/domestic-sales-trend.aspx> (Accessed 16 August 2014).

Sonpo Soken (General Insurance Institute of Japan) (2013) *Ajia syokoku ni okeru songai hoken shijyou, syoseido no gaiyou ni tuite (Insurance Market and System in Asian Countries)* (in Japanese). Tokyo: Sonpo Soken.

South African Revenue Service (SARS) (2009) *Valued added Tax. Guide for Motor Dealers.* Online. Available: <http://www.sars.gov.za/AllDocs/OpsDocs/Guides/ LAPD-VAT-G09%20-%20VAT%20420%20Guide%20for%20Motor%20Dealers%20- %20External%20Guide.pdf> (Accessed 18 November 2013).

South African Revenue Service (SARS) (2010) *Environmental Levy on Carbon Dioxide Emissions of New Motor Vehicles Manufactured in the Republic.* Online. Available: <http://www.sars.gov.za/ClientSegments/Customs-Excise/Excise/Environmental-Levy-Products/Pages/Motor-vehicle-CO2-emmision-levy.aspx> (Accessed 1 October 2013).

South African Revenue Service (SARS) (2013) *Motor Vehicle CO$_2$ Emissions*. Online. Available: <http://www.sars.gov.za/ClientSegments/Customs-Excise/Excise/Environmental-Levy-Products/Pages/Motor-vehicle-CO2-emmision-levy.aspx> (Accessed 31 October 2013).

Stechdaub, R. (2001) "Inspection and maintenance in Indonesia," in UNESCAP and JARI (eds.), *GITE Regional Workshop on Inspection and Maintenance Policy in Asia*, Bangkok, Thailand, 10–12 December.

Suwanvithaya, C. (2013) "Tax and insurance in Thailand." Email (Personal communication, 26 November 2013).

Suweden I.N. (2013) "Tax and insurance in Indonesia." Email (Personal communication, 25 November 2013).

Tamang, J., 2011. *Philippines Energy Outlook 2009–2030*. Manila: DOE.

Techawiboonwong, A. (2014) "Insurance in Thailand." Email (Personal communication, 13 January 2014).

Thailand Automotive Institute (TAI) (2014) "Ongoing and future initiative on emission and fuel economy regulations with consideration for vehicle tax." Email (Personal communication, 20 August 2014).

Transport Department Government of Kartanaka (TDGK) (2010) *Karnataka motor vehicle taxation Act 2010*. Bangalore: Department of Transport.

Transport Research Wing (TRW) (2009) *Road transport yearbook (2006–2007)*. Delhi: Ministry of Shipping, Road Transport and Highway, Government of India.

UNESCAP and JARI (2001) "Conclusions and recommendations of workshop," in UNESCAP and JARI (Eds.) *GITE Regional Workshop on Inspection and Maintenance Policy in Asia*, Bangkok, Thailand, 10–12 December.

Wagner, D.V., An, F. and Wang, C. (2009) "Structure and impacts of fuel economy standards for passenger cars in China," *Energy Policy*, 37: 3804–11.

Wangwongwatana, S. (2007) "Project of vehicle emission reduction in Thailand," in *JARI Thailand Roundtable*. Bangkok, Thailand, 30 March. Tsukuba: JARI.

World Bank (2014) *DATA: Passenger Car Per 1000 People*. Online. Available: <http://data.worldbank.org/indicator/IS.VEH.PCAR.P3?page=1> (Accessed 16 August 2014).

Xu, W. (2013) "Taxation in China." Email (Personal communication, 26 November 2013).

Yan, Y. (2010) "Air pollution reduction from vehicles in Beijing," in *JARI China Roundtable 2010*. Beijing, China, 15 November. Tsukuba: JARI.

Yasin, S.B. and Zainudin, M.B. (2011) "Emission regulation in Malaysia." Interview (1 December 2011).

Yuan, Y. (2005) *Inspection Standard for Vehicle In-Use*. August, p. 10. Online. Available: <http://www.cpcb.nic.in/oldwebsite/News%20Letters/Archives/Inspection%20Maintenance/ch6-IM.html> (Accessed 7 April 2011).

Zakaria, M. (2013) *Cost Benefit Analysis on Fuel Quality and Fuel Economy Initiative*. Online. Available: <http://www.globalfueleconomy.org/Documents/muhamad_zakaria.pdf> (Accessed 17 August 2014).

9 CO_2 emission reductions from hybrid vehicle use

An analysis of Japan's used car market data

Kazuyuki Iwata and Shigeru Matsumoto

9.1. Introduction

Reducing greenhouse gas (GHG) emissions from the transportation sector forms an important part of the climate change policy agenda. For the last several years, countries have significantly tightened fuel economy standards for motor vehicles. For example, on 28 August 2012, the Obama administration introduced a standard that required automakers to nearly double the average fuel economy of new cars and trucks by 2025 (Vlasic 2012). On 23 April 2009, the European Parliament and the Council of the European Union approved regulations setting a target of 130 g/km for average emissions from new cars, which will be gradually phased in by 2015 (Global Fuel Economy Initiative 2013).

Hybrid vehicles (HVs) are considered to be a cost-effective, practical solution for reducing GHG emissions from the transportation sector. In recent years, many countries have implemented rebate programs to stimulate HV sales.[1] For instance, federal, state, and local governments in Canada and the United States have implemented sales tax or cash rebate programs for HVs. Empirical studies report that these programs have led to a large increase in the market share of HVs (Chandra, Gulati, and Kandlikar 2010; Gallagher and Muehlegger 2011).

The Japanese government implemented a similar rebate program for HVs. Figure 9.1 presents the change in the sales share of HVs among all passenger vehicles in the Japanese market. The figure shows a strong upward trend in the market penetration of HVs.[2] Although the sales share was about 5 percent in 2008, it has exceeded 30 percent since 2011. Gasoline vehicles are no longer the top-selling vehicle type in the Japanese market; HVs took over the top position as of May 2009. According to estimates by the Next Generation Vehicle Promotion Center (2013), HV ownership was 2,012,559 in 2011.[3] Since the total number of owned vehicles in 2011 was 40,135,102, HVs are estimated to comprise 5 percent of currently owned vehicles.

The Japanese government implemented two rebate programs to accelerate the market penetration of eco-friendly vehicles. The government spent 589.9 billion yen and 300 billion yen ($4.92 billion and $2.5 billion, respectively) for the first and the second rebate programs. Figure 9.1 also presents the timeline of key events in the two rebate programs. Consumers who purchased an

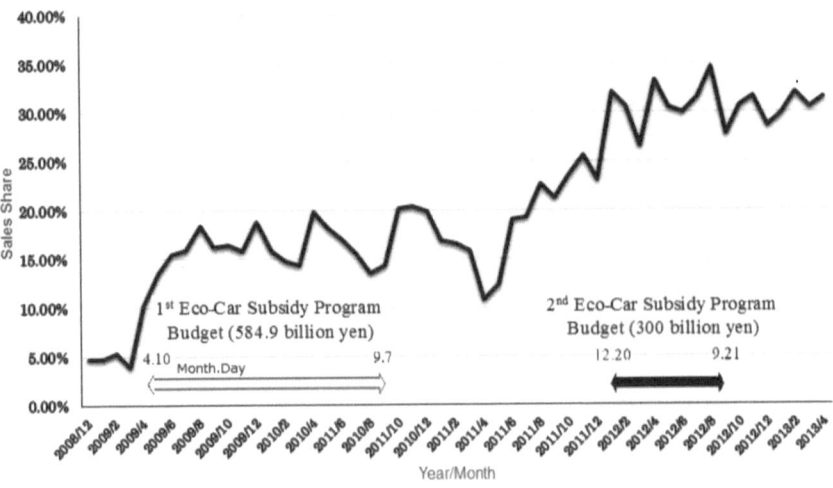

Figure 9.1 Sales share of HVs among passenger vehicles

eco-friendly vehicle during the program period could obtain a cash rebate depending on the vehicle's fuel economy criteria.[4] The first rebate program commenced on 10 April 2009. Immediately after its implementation, the sales share of HVs increased from 5 to 15 percent. After the completion of the first rebate program on 7 September 2010, the market share of HVs decreased to 10 percent. However, it jumped to 30 percent after the second rebate program was introduced on 20 December 2011. The sales share of HVs has remained at around 30 percent, even after the completion of the program on 21 September 2012.

Despite the rapid market penetration of HVs, the driving behavior of HV owners has not yet been examined carefully. The majority of previous studies assumed that the driving behavior of HV owners was similar to that of conventional gasoline vehicle owners and then estimated the potential reduction of carbon dioxide (CO_2) emissions. In fact, the rebate amount as well as the tax reduction rate are based solely on the fuel economy of vehicles.

In this study, we analyze vehicle usage by employing Japan's used car market data and evaluate the effectiveness of the rebate program for HVs. Specifically, we examine (1) whether there was a difference between the mileage traveled by HVs and their competitor vehicles, (2) how much CO_2 HV drivers saved, and (3) whether the rebate program for HVs was cost-effective as a CO_2 reduction measure.

The remainder of this chapter is structured as follows. We provide background information in Section 9.2. We summarize the used car market data in Section 9.3. We compare the mileages traveled across vehicle models in Section 9.4. We report that an HV owner drove a longer distance than a conventional gasoline vehicle owner. In Section 9.5, we examine how much CO_2 HV drivers saved.

We show that despite the longer mileage traveled, HV owners emitted less CO_2 owing to better fuel economy. In Section 9.6, we evaluate the cost-effectiveness of the rebate program. We argue that the rebate amount for an HV exceeds the social benefit from CO_2 reduction. Section 9.7 concludes the chapter and discusses policy implications.

9.2. Background

Although an HV is often considered to be a new, emerging technology, it has a long history. Ferdinand Porsche was reportedly the first to design and build an HV as early as 1896. He installed an electric motor as an auxiliary power source to increase the horsepower and speed of the gasoline vehicle. However, technological innovations in internal combustion engines created sufficient horsepower and speed for gasoline vehicles. The idea of an HV was then abandoned, and it has only recently returned in vogue (Denki Jidosha Dot Com 2013).

Improvements in battery and motor technologies and the growing concern about global warming have helped HVs regain popularity in recent years. Especially since Toyota's success in mass producing the Prius, HVs have become an affordable option for consumers.

To further accelerate the market penetration of HVs, many countries reduced vehicle-related taxes, and in recent years some have even begun providing cash rebates for HVs. The Japanese government also reduced vehicle-related taxes and provided cash rebates for HVs.

Besides fuel taxes, three kinds of vehicle-related taxes (acquisition, weight, and automobile taxes) were imposed in Japan. Table 9.1 compares the vehicle-related taxes for regular gasoline and HV vehicles. Suppose that a consumer

Table 9.1 Tax reduction for hybrid vehicles before 1 May 2012 (First three years: weight = 1.3 ton, engine size = 1,800 cc, acquisition price = 2,500,000 yen)[a]

	Regular Gasoline Vehicle	*Hybrid Vehicle*
Acquisition tax[b] (local tax)	125,000 yen	0 yen
Weight tax[c] (national tax)	45,000 yen	0 yen
Automobile tax[d] (local tax)	118,500 yen	99,000 yen
Total	288,500 yen	99,000 yen

a $1 is approximately 120 yen.

b Acquisition price × 5%.

c The annual tax rate is 5,000 yen/0.5 t. Hence, the required tax for the first three years is 5,000 × (1.5 t/0.5 t) × 3 = 45,000 yen.

d The annual tax rate is 39,500 yen. Hence, the required tax is 39,500 × 3 = 118,500 yen. The initial-year tax for HVs is lowered to 20,000.

purchases a vehicle with a weight and an engine size of 1.3 ton and 1,800 cc, respectively, at 2,500,000 yen ($20,833.33). If s/he decides to purchase an HV, then s/he could save 189,500 yen ($1,579.12) on vehicle-related taxes for the first three years. In addition to the benefit of this tax reduction, s/he could obtain a cash rebate of 100,000 yen ($833.33) if s/he purchased it during one of the two rebate program periods.

As discussed in Section 9.1, the rebate programs rapidly increased HV sales in the Japanese market. According to the Japan Automobile Dealers Association (2013), 3,014,651 new domestic vehicles were sold in the Japanese market in 2012, of which 899,155 were HVs. Thus, the sales share of HVs among passenger vehicles is estimated to be about 29.8 percent. Although only two HVs (the Prius and the Insight) were available during the downturn in HV sales, the number of HV varieties has increased in recent years. In 2013, 33 varieties of HVs were sold in the Japanese market.

The market share of HVs in the global market is much smaller than that in the Japanese market. For example, Toyota (2013) reported that while worldwide cumulative sales of HVs exceeded 5.13 million by March 2013, 2.32 million (45.3 percent) of these could be attributed to the Japanese market alone.

Nonetheless, HV sales outside Japan are forecast to increase in the near future. For instance, Fuji Keizai (2013) estimated that the market share of HVs in Europe and North America will rise to 17.8 percent and 12.2 percent by 2025, respectively. Therefore, it is worthwhile to examine HV use in Japan.

The most attractive feature of an HV is its superior fuel economy. According to the 10.15 mode data released by the Ministry of Land, Infrastructure, Transport and Tourism (2013), the fuel economy of the Prius is 38.0 km/L, while the corresponding average of other gasoline vehicles is about 20.0 km/L. Thus, a large gap exists between the fuel economies of HVs and gasoline vehicles.[5] The main attraction of purchasing an HV is the savings it provides in fuel costs (Gulliver 2006). Past studies, such as Gallagher and Muehlegger (2011), have reported that rising gasoline prices are associated with greater HV sales.

However, the HV's high acquisition price may also be looked upon as its one weakness. For instance, the Honda Civic is manufactured in both gasoline and hybrid models. The price difference between the two models is about 300,000 yen ($2,500). Since the difference in fuel economy between these two models is about 14 km/L (= 31 – 17) and the difference in vehicle-related taxes is about 130,000 yen ($1,083.33), the driver has to drive more than 42,662 km before the high acquisition cost is compensated.[6] As we report later, the typical average mileage is less than 3,000 km per year. Thus, it takes more than 10 years to make up for the HV's high acquisition price.

9.3. Data

We obtain our primary data from Proto Corporation (2010), which publishes one of the most popular used car magazines, *Goo*. It also releases up-to-date information on a website called "Goo-net."

The Goo data cover all regions in Japan. We focus on the data for March and September, released between 2008 and 2010. March and September are the two peak months in the Japanese used car market. For example, 4,015,909 used vehicles were registered in 2012. In the same year, 528,192 and 306,304 used vehicles were registered in March and September, respectively. The number of used cars in the Goo data for March 2012 is 340,301, while that for September 2012 is 352,643. Therefore, the data coverage is reasonably high.

The data include a variety of information such as the price, mileage, and detailed specifications of the used cars. We used this data to conduct the following three analyses.

In the first stage of the analysis, we compared the mileages traveled across vehicle models and then identified the vehicle characteristics associated with a particular mileage. We present the results of the first stage of the analysis in Section 9.4.

Parry, Walls, and Harrington (2007) estimated the monetary values of the externalities associated with vehicle use in the United States. Kanemoto (2007) and Koyama and Kishimoto (2001) used a similar analytical approach to estimate the monetary values of externalities in Japan. In the second stage of the analysis, we combined their external costs with the mileage data and compared the magnitudes of externalities across vehicle models.

As mentioned before, the government spent a substantial amount of money on the rebate programs. The sales data in Figure 9.1 reveals that the rebate program has succeeded in promoting HVs. However, no examination has been conducted on whether the subsidy program was cost-effective as a CO_2 mitigation strategy. Thus, we included vehicle-related taxes and rebates in the second stage of the analysis and evaluated the cost-effectiveness of the rebate program for HVs.

9.4. Mileage analysis

9.4.1. Comparing mileages of HVs and competitor vehicles

In this section, we compare the mileages traveled by the Prius and its competitor gasoline vehicles. We chose eight varieties of vehicles (Mazda Axela Sports, Peugeot 207, Nissan Note, Honda Fit, Toyota Wish, Toyota Vitz, Volkswagen Golf, and Toyota Auris) as the competitor vehicles based on Autoc One's monthly report of March 2010 (Autoc One 2012). Autoc One provides an intermediary service on its website. A consumer is allowed to submit requests for quotation (RFQ) on varieties of vehicles. The monthly report summarizes the RFQ data.

Table 9.2 compares the average mileages across vehicles. The first column presents the vehicle models; the second column shows the average mileage of the models. We used the data released in March 2010 and calculated the average mileages of the models manufactured in 2007. The table shows that the average mileage of the Prius is 35,769.5 km for the first three years. It far

Table 9.2 Mileage comparison between HV and competitor vehicles[a]

		Average Mileage (km)	Variance (km)	Number of Samples	t-value[b]
Prius	$\bar{y}_h$	35,769.5	360,830,969.6	243	
Competitor vehicles					
Mazda Axela Sports		22,756.4	150,316,516.8	78	7.04***
Peugeot 207	$\bar{y}_g$	17,280.0	73,876,666.7	25	8.77***
Nissan Note		29,825.8	303,918,650.4	178	3.33***
Honda Fit		27,000.0	247,587,412.6	287	5.72***
Toyota Wish		30,894.2	238,095,039.0	208	3.01***
Toyota Vitz		25,493.2	285,643,206.4	386	6.89***
Volkswagen Golf		28,079.5	182,280,956.1	88	4.08***
Toyota Auris		24,014.2	211,023,606.4	211	7.46***

a The data refer to models manufactured in 2007 and are sourced from the Goo report released in March 2010.
b Welch's t-test ($\bar{y}_h = \bar{y}_g$) was conducted. *, **, and *** indicate significance at the 10%, 5%, and 1% levels, respectively.

exceeds the corresponding mileages for the competitor gasoline vehicles. The t-statistics in the last column confirm that Prius drivers traveled much longer distances than the other drivers did.

9.4.2. Determinants of mileage

To identify which vehicle characteristics are correlated with mileage, we included an additional eight vehicle models into our dataset. Then, we estimated the following regression model:

$$\ln\left(AMTD_{igt}\right) = \alpha + \sum_j \beta^j \ln x_{igt}^j + \gamma Prius + \sum_t \delta_t D_t + \sum_r \delta_r D_r + \varepsilon_{igt} \qquad (9.1)$$

where i, g, t, and r stand for the vehicle number, vehicle model, vehicle model year, and sales region, respectively. The variable of $AMTD_{igt}$ is the average monthly travel distance (AMTD).[7] The vector of X_{igt} includes the engine size, weight, holding period, and fuel economy. The descriptive statistics of these variables are presented in Table 9.3. The variable of D_t is a year-specific dummy variable, while the variable of D_r is a sales region–specific dummy variable.

The estimation results of Equation 9.1 are presented in Table 9.4. Since there is a strong correlation between engine size and weight, we evaluated the effects of the two variables separately. Model 1 includes the engine size variable only, while Model 2 includes the weight variable only. We calculated the adjusted

Table 9.3 Determinants of mileage: descriptive statistics (N = 29,950)

Variable	Unit	Average	Standard Deviation	Min	Max	Number of Samples
Monthly travel distance	km	692.44	483.37	3.07E–05	5,000	29,950
Engine size	cc	1,401.29	252.19	996	3,188	29,950
Weight	kg	1,081.82	142.99	865	1,580	29,950
Fuel economy	km/L	19.91	4.60	9.1	38	29,950

weight by dividing vehicle weight by engine size. In Model 3, we included it together with the engine size variable. In Model 4, we included the Prius dummy. All four models use the robust sandwich estimator. The discussion of the estimation results of the first three models appears below.

Although we observe differences in coefficient values across the models, the table shows that all variables have a similar explanatory power on AMTD. First, the table shows that the AMTD of a vehicle with a large engine is shorter. According to the estimation, a 1 percent increase in engine size increases the AMTD by 0.462 to 0.856 percent. We also obtain a positive sign for the weight variables. These results imply that people with heavier cars tend to drive longer distances.

The coefficient of fuel economy is positive and statistically significant in all three models. The results show that a 1 percent increase in fuel economy raises the AMTD by 0.351 to 0.383 percent. This implies that drivers with vehicles having good mileages tend to drive longer distances.

For illustrative purposes, consider a person with a vehicle having a fuel economy of 20 km/L, which travels 660 km per month. In this example, s/he consumes 33 L of gasoline per month. Suppose that the fuel economy of the vehicle increases by 10 percent. Will the driver reduce her/his gasoline consumption by 10 percent? Previous studies have provided empirical evidence that a 10 percent increase in energy efficiency reduces gasoline use by less than 10 percent (Sorrell, Dimitropoulos, and Sommerville 2009). This phenomenon is called the "rebound effect." According to our estimation, the AMTD increases by 3.51 to 3.83 percent when the fuel economy increases by 10 percent. Consequently, the driver is expected to travel 683.17 km as a result of the fuel economy improvement. The driver will consume 31.05 L (= 683.17 km ÷ 22 km/L) of gasoline, reducing gasoline consumption by only 6.5 percent.

Mizobuchi (2011) estimated the rebound effect among Japanese drivers and found that the size of the rebound effect was about 18 percent. In this chapter, however, we find evidence of a much larger effect. Perhaps the difference between our results and Mizobuchi's (2011) can be attributed to the fact that a person who purchases a vehicle with a good mileage has a higher travel demand than one who purchases a vehicle without it. The present analysis does not control

Table 9.4 Determinants of mileage (N = 29,950)

Variable	Model 1 Coefficient	Model 1 Std. Error[a]	Model 2 Coefficient	Model 2 Std. Error	Model 3 Coefficient	Model 3 Std. Error	Model 4 Coefficient	Model 4 Std. Error
Constant	1.039***	0.444	-1.674***	0.460	-1.443***	0.477	-3.742***	0.867
Engine size	0.462***	0.048		0.056	0.856***	0.056	1.107***	0.094
Weight			0.868***	0.056				
Adjusted Weight					1.034***	0.089	1.263***	0.107
Automatic dummy	0.141***	0.035	0.084**	0.035	0.071**	0.036	0.054	0.035
Period of possession	-0.412E-04***	0.431E-05	-0.411E-04***	0.430E-05	-0.411E-04***	0.430E-05	-0.411E-04***	0.430E-05
Fuel economy	0.366***	0.042	0.383***	0.037	0.351***	0.042	0.540***	0.077
Prius dummy							-0.201***	0.064
2004 model	0.851***	0.040	0.863***	0.040	0.866***	0.040	0.888***	0.041
2005 model	0.873***	0.040	0.878***	0.040	0.878***	0.040	0.892***	0.041
2006 model	0.818***	0.041	0.816***	0.041	0.816***	0.041	0.825***	0.041
2007 model	0.620***	0.041	0.623***	0.041	0.622***	0.041	0.628***	0.041
2008 model	0.134***	0.044	0.144***	0.044	0.145***	0.044	0.149***	0.044
Adjusted R^2	0.1418		0.1451		0.1452		0.1455	

a Standard error.

*, **, and *** indicate significance at the 10%, 5%, and 1% levels, respectively.

for individual characteristics. Therefore, we may estimate the difference in travel demand rather than the rebound effect. To estimate the latter, we need to examine whether a driver increased her/his travel distance after purchasing a vehicle with a good mileage.

We obtained a negative sign for the Prius dummy variable in the last model. This result implies that the better mileage of the Prius can be explained by its high fuel economy. Suppose that the fuel economy of the Prius exceeds that of the other vehicles by 30 percent. Then, we expect that the AMTD of the Prius would exceed that of the other vehicles by 16 percent [(0.540×30) – 0.201]. This explains the difference in mileage between the Prius and the other gasoline vehicles.

9.5. External cost analysis

9.5.1. *Assumptions about external costs*

A variety of negative externalities are generated by automobile use. Kanemoto (2007) and Koyama and Kishimoto (2001) estimated the monetary values of negative externalities in Japan. Table 9.5 summarizes their estimation results.

Table 9.5 Summary of external cost (unit: yen/km)[a]

	Median Value		Lower Bound – Upper Bound	
	Koyama and Kishimoto (2001)	*Kanemoto (2007)*	*Koyama and Kishimoto (2001)*	*Kanemoto (2007)*
Fuel-related costs				
Global warming	2.2	2.0	0.05–17.7	0.3–3.4
Oil dependency	–	0.5	–	0–1.3
Sum	2.2	2.5	0.05–17.7	0.3–4.7
Mileage-related costs				
Local pollution	1.8	1.1	1.1–2.6	0.1–3.2
Congestion	7.1	2.5	Not reported	0.7–4.8
Accidents	7.3	7.0	2.0–14.6	0–36
Noise	3.6	Not estimated	1.3–5.2	Not estimated
Road damage	Not estimated	0.1	Not estimated	Not estimated
Cost of infrastructure	7.0	Not estimated	Not reported	Not estimated
Sum	26.8	10.7	4.4–22.4	Not reported
Grand Total	29.0	13.2	4.45–40.1	1.2–48.8

a $1 is approximately 120 yen.

The table shows a wide variation in the values of the externalities. For instance, Koyama and Kishimoto (2001) found that the median value of the marginal cost of CO_2 is 34,408 yen/ton ($286.73/ton). However, according to the previous literature, the marginal cost ranges from 850 yen/ton ($7.08/ton) to 274,329 yen/ton ($2,286.08/ton) (Parry, Walls, and Harrington 2007). This implies that careful consideration is necessary when gauging the values of the externalities.

Both studies, however, reported that mileage-related costs are much larger than fuel-related costs. This means that only a small fraction of externality problems can be resolved by fuel economy improvements.

9.5.2. Savings realized on external costs

By combining the external costs in Table 9.5 with the mileage data in Table 9.2, we compared the external costs across vehicle models. The results are presented in Table 9.6. The average mileage of HV (Prius) owners for the first three years is 35,770 km, while the corresponding values for the competitive vehicles range from 17,280 km to 30,894 km. Therefore, HV (Prius) owners drive their cars more than conventional car owners. The table shows that despite the higher

Table 9.6 Comparison of external costs between vehicle models (first three years)

Model (Unit)	Average Mileage (km)	Average CO_2 Emissions[a] (kg/Vehicle)	Costs of Global Warming[b] (yen)[d]	Mileage-Related Costs[c] (yen)[d]	Total External Cost (yen)[d]
Prius	35,770	2,325.0	23,250	382,734	405,984
Competitor vehicles					
Mazda Axela Sports	22,756	3,433.1	34,331	243,494	277,824
Peugeot 207	17,280	3,464.8	34,648	184,896	219,544
Nissan Note	29,826	3,626.9	36,269	319,137	355,405
Honda Fit	27,000	2,768.7	27,687	288,900	316,587
Toyota Wish	30,894	5,066.1	50,661	330,568	381,230
Toyota Vitz	25,493	2,715.2	27,152	272,778	299,930
Volkswagen Golf	28,080	5,024.8	50,248	300,451	350,699
Toyota Auris	24,014	3,252.9	32,529	256,952	289,481

a Based on the 10.15 mode data, we estimated the CO_2 emissions of each vehicle. We aggregated all CO_2 emissions and divided that value by the number of vehicles to estimate the average CO_2 of each vehicle model.

b The cost of CO_2 is assumed to be 10,000 yen/ton.

c Mileage-related external costs are assumed to be 10.7 yen/km. We assumed that the mileage-related external costs of HVs are the same as those of regular gasoline vehicles.

d $1 is approximately 120 yen.

mileage, HV owners emit lower amounts of CO_2 owing to better fuel economy. Therefore, we can state that HVs can mitigate global warming. If the cost of CO_2 is 10,000 yen/ton ($83.33/ton), we estimate that each HV owner saved approximately 11,000 yen to 27,000 yen ($91.67 to $225) in social costs for the first three years.

If the other mileage-related external costs listed in Table 9.5 are included in the analysis, then the economic benefit of HVs with respect to conventional vehicles disappears. Following Kanemoto (2007), for our estimation in Table 9.6, we assumed that the median mileage-related external cost is 10.7 yen/km ($0.09/km). By multiplying this value with the mileage, we estimated that the mileage-related external cost of the Prius is 382,734 yen ($3,189.45). This external cost is much larger than the external costs of the other vehicles.

Thus, although hybrid technology mitigates global warming, it will not solve the other externality problems. The mileage-related costs are much higher than the fuel-related costs. Since we expect the market penetration of HVs to deepen in the near future, it will be necessary to implement a vehicle miles traveled (VMT) tax.

9.6. Cost-effectiveness of the rebate programs

The analysis in Section 9.5 confirms that CO_2 emissions from the transportation sector can be reduced by replacing regular gasoline cars with HVs. Therefore, it is rational for the government to promote HVs. Nonetheless, a significant amount of money has been spent on their promotion. Given the above empirical results, can we state that the rebate programs have been cost-effective? In this section, we discuss the cost-effectiveness of the rebate programs.

In Table 9.7, we estimate the optimum rebate for the Prius against the other vehicle models. For example, the Mazda Axela Sports emits 3,433.1 kg of CO_2 for the first three years. In contrast, the corresponding number for the Prius is 2,325.0 kg. The difference in CO_2 emissions for the first three years is 1,108.1 kg. If the cost of CO_2 is 5,000 yen/ton ($41.67/ton), then the optimum rebate or tax difference will be 5,540.5 yen ($46.17/ton). If the cost of CO_2 is 10,000 yen/ton ($83.33/ton), the corresponding value will be 22,162.0 yen ($184.68).

As shown in the bottom row in Table 9.7, the average difference in CO_2 emissions is 1,344.1 kg for the first three years. Therefore, the optimum rebate or tax difference will be 6,720.3 yen ($56.00) if the cost of CO_2 is 5,000 yen/ton ($41.67/ton), and 26,881.3 yen ($224.01) for a cost of 20,000 yen/ton ($166.67/ton). Although the price of carbon is widely debated, it would be reasonable to assume that it was less than 20,000 yen/ton ($166.67/ton) when the rebate programs were implemented in Japan.[8] Since the rebate amount for HVs was larger by 31,250 yen ($260.42) than the corresponding values for the competitor vehicles, we conclude that HV owners received a larger rebate compared to their contribution of CO_2 reductions. Therefore, the rebates were larger than they should have been.

Table 9.7 Optimum rebate for HVs

Estimated CO_2 Emissions (kg/Vehicle)				*Cost of CO_2 (yen/ton)*[a]			
Competitor Vehicles		*Prius*	*Difference*	*5,000*	*10,000*	*20,000*	*30,000*
Mazda Axela Sports	3,433.1	2,325.0	1,108.1	5,540.5	22,162.0	33,243.0	55,405.0
Peugeot 207	3,464.8	2,325.0	1,139.8	5,699.0	22,796.0	34,194.0	56,990.0
Nissan Note	3,626.9	2,325.0	1,301.9	6,509.5	26,038.0	39,057.0	65,095.0
Honda Fit	2,768.7	2,325.0	443.7	2,218.5	8,874.0	13,311.0	22,185.0
Toyota Wish	5,066.1	2,325.0	2,741.1	13,705.5	54,822.0	82,233.0	137,055.0
Toyota Vitz/	2,715.2	2,325.0	390.2	1,951.0	7,804.0	11,706.0	19,510.0
Volkswagen Golf	5,024.8	2,325.0	2,699.8	13,499.0	53,996.0	80,994.0	134,990.0
Toyota Auris	3,252.9	2,325.0	927.9	4,639.5	18,558.0	27,837.0	46,395.0
Average			1,344.1	6,720.3	26,881.3	40,321.9	67,203.1

a $1 is approximately 120 yen.

9.7. Conclusion

In this chapter, we analyzed the used car market data in Japan and examined whether a difference exists between HV owners and gasoline vehicle owners in terms of travel demand. The mileage data of used vehicles demonstrates that HV owners drive a longer distance than gasoline vehicle owners. This difference can be explained by the difference in their respective fuel economies. Despite the higher mileage, it is estimated that CO_2 emissions from HVs are less than those from gasoline vehicles. Therefore, it is natural to expect that the promotion of HVs will mitigate global warming.

Although many governments have implemented various rebate programs for HVs, such programs may require several adjustments. Considering the carbon cost applied for our analyses, the rebate and tax reduction offered for HV purchases in Japan are too large, and thus incentives provided to HVs need to be reviewed.

Although GHG emissions will be reduced by the market penetration of HVs, the other externality problems will not be resolved. The mileage-related external cost is much larger than the cost of GHG mitigation. The current subsidy program provides the privilege to drivers traveling longer distances. Thus, a mileage-related charge is required to solve the other externality problems associated with HV use.

Acknowledgments

An earlier version of this chapter was presented at the 2nd Symposium of the European Association for Research in Transportation and the 12th International Conference of the Japan Economic Policy Association. We thank Akira Maeda and Toru Morotomi for their helpful comments.

Notes

1 Throughout this chapter, we use the words "subsidy" and "rebate" interchangeably.
2 Only regular and compact passenger vehicles are included in this discussion. Light motor vehicles are excluded.
3 Only regular and compact passenger vehicles are included.
4 Japan's fuel economy criteria are based on vehicle weight. A strict criteria set is applied for light vehicles and a lax one for heavy vehicles.
5 The 10.15 mode is the official fuel economy test cycle for new cars and is expressed in km/L (kilometers per liter). Carried out on a dyno rig, the 10.15 mode cycle is a series of 25 tests that cover idling, acceleration, steady running, and deceleration and that simulate typical urban and/or expressway driving patterns (Japan Automobile Manufacturers Association 2009). Critics suggest that the 10.15 mode cycle overestimates the fuel economy of vehicles. Thus, fuel economy based on the 10.15 mode cycle is estimated to be much higher than the actual fuel economy. However, we use the 10.15 mode cycle data in this chapter since the other data do not cover as many vehicles.
6 Another weakness could be the battery replacement cost. However, most vehicle reviews have reported that battery replacement is not necessary during product usage.

7   The data include information about the next car inspection day. Using this information and the model year of a vehicle, we gauged the date that the vehicle was first driven (i.e. the driving starting date). Then, we calculated the difference between the advertisement date and the driving starting date and defined this difference as the holding period. Finally, we divided the mileage by the holding period and used the resulting value as the average monthly travel distance (AMTD).

8   For instance, the Ministry of Environment (2013) estimates the marginal abatement cost curve and shows that more than 40 million tons of CO_2 can be mitigated below 20,000 yen/ton in Japan.

References

Autoc One KK. (2012) 'Mach 2010 Prius competition analysis', Available: <http://autoc-one.jp> (accessed 10 August 2012) (in Japanese).

A. Chandra, S. Gulati, and M. Kandlikar. (2010) 'Green drivers or free riders? An analysis of tax rebates for hybrid vehicles', *Journal of Environmental Economics and Management* 60, 78–93.

Denki Jidosha Dot Com. (2013) 'History of hybrid vehicles', Available: <http://ele-car.com> (accessed 15 July 2013) (in Japanese).

Fuji Keizai Management Co. Ltd. (2013) 'HV, PHV and EV world market analysis: prediction of 2030 world market', Available: <https://www/fuji-keizai.co.jp> (accessed 15 July 2013) (in Japanese).

K. Gallagher and E. Muehlegger. (2011) 'Giving green to get green? Incentives and consumer adoption of hybrid vehicle technology', *Journal of Environmental Economics and Management* 61, 1–15.

Global Fuel Economy Initiative. (2013) 'The European Union automotive fuel economy policy', Available: <http://www.unep.org/transport/gfei/autotool/basic.asp> (accessed 25 February 2013).

Gulliver International Co., Ltd. (2006) 'Car life research, 2006 June', Available: <http://221616.com/gulliver/wp-content/uploads/2009/09/etc_010.pdf> (accessed 1 August 2013).

Japan Automobile Dealers Association. (2013) 'Statistical information', Available: <http://www.jada.or.jp> (accessed 16 July 2013).

Japan Automobile Manufacturers Association. (2009) 'From 10.15 to JC08: Japan's new economy era. News from JAMA', Available: <http://www.jama-english.jp> (accessed 12 March 2013).

Y. Kanemoto. (2007) 'Economic analysis of road taxes and earmarking', Chapter 1, 1–44, The Japan Research Center for Transport Policy, Series A-430 (in Japanese).

S. Koyama and A. Kishimoto. (2001) 'External costs of road transport in Japan', *Transport Policy Studies* 4, 19–30 (in Japanese).

Ministry of the Environment. (2013) 'Policy measures to maximize CO_2 abatement potential', Available: <http://www.env.go.jp/earth/er-potential/05/mat03.pdf> (accessed 10 January 2014) (in Japanese).

Ministry of Land, Infrastructure, Transport and Tourism of Japan. (2013) 'Automobile fuel economy list: March 2012', Available: <http://ele-car.com> (accessed 15 July 2013) (in Japanese).

K. Mizobuchi. (2011) 'Rebound effect for passenger vehicles: estimations with micro panel data', *Kankyo Keizai Seisaku Kenkyu* (*Environmental Economics Policy Studies*) 4, 32–40 (in Japanese).

Next Generation Vehicle Promotion Center. (2013) 'Statistics of the number of electric vehicles owned (estimation)', Available: <https://www.cev-pc.or.jp/> (accessed 22 July 2013) (in Japanese).

I. Parry, M. Walls and W. Harrington. (2007) 'Automobile externalities and policies', *Journal of Economic Literature* 45, 373–99.

Proto Corporation. (2010) 'Goo-net', Available: <http://www.goo-net.com/index.html> (accessed 10 May 2010) (in Japanese).

S. Sorrell, J. Dimitropoulos, and M. Sommerville. (2009). 'Empirical estimates of the direct rebound effect: A review', *Energy Policy* 37, 1356–71.

Toyota. (2013) 'Cumulative sales of hybrid vehicles exceeded 5,000,000', Available: <http://www.toyota.co.jp> (accessed 17 April 2013).

B. Vlasic. (2012) 'U.S. sets higher fuel efficiency standards', *New York Times*. Available: <http://www.nytimes.com/2012/08/29/business/energy-environment/obama-unveils-tighter-fuel-efficiency-standards.html> (accessed 25 February 2013).

Part IV

Environmental subsidies to consumers

Multiple points

10 Subsidy for eco-friendly houses

Can it be an ultimate solution?

Kazuyuki Iwata

10.1. Introduction

Various approaches to reducing household greenhouse gas (GHG) emissions are available to policymakers, including promoting energy-efficient appliances and automobiles (e.g. Parts II and III in this volume), encouraging energy-saving behaviors (e.g. Costanzo et al. 1986), and constructing energy-efficient houses. In 2011, approximately 66 percent of energy consumption in the Japanese residential sector was attributable to home heating and cooling, water heating, and the use of kitchen appliances (Energy Conservation Center Japan [ECCJ] 2013). The remainder was primarily attributable to automobiles. Because Parts II and III of this book address energy-efficient appliances and automobiles, respectively, this chapter focuses on energy-efficient houses. This chapter examines the effectiveness of a large-scale program implemented in Japan that subsidized the construction of energy-efficient "eco-friendly" houses. The goal of the program was to substitute environmentally-friendly housing materials for conventional ones, thereby reducing household GHG emissions.

The Ministry of the Environment (MOE) defines an "eco-friendly house" as a house with the following two characteristics:[1] (1) high environmental performance, such as low thermal conductivity, and (2) the use of natural or renewable energy sources, such as solar photovoltaic electricity. The former relates to housing materials, the area of focus here, whereas the latter relates to the promotion of sustainable energy. Therefore, in this chapter, only houses constructed with energy-efficient materials are considered eco-friendly.

Household energy consumption in Japan, including electricity, gas, and kerosene, has increased considerably with economic growth. Detailed historical energy consumption data for the Japanese residential sector show that space heating and cooling accounts for a consistently large portion of consumption, approximately 26.8 percent, 29.0 percent, and 28.9 percent in 1990, 2000, and 2011, respectively (ECCJ 2013). This is attributable to Japan's wide seasonal variation in temperature as well as its geographical characteristics. Many households use air conditioners to reduce the indoor temperature in the summer and raise it during the winter, although to what extent households use air conditioners depends on their specific locale. The number of air conditioners

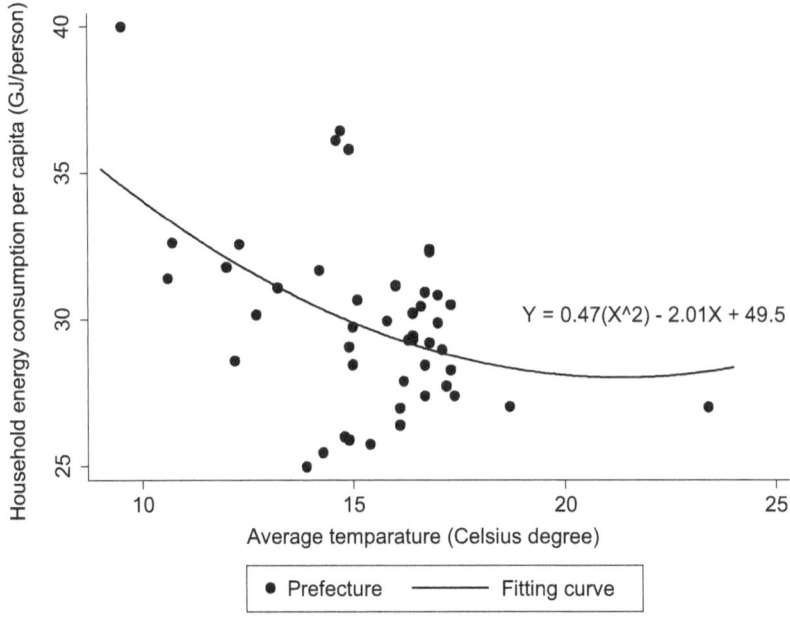

$$Y = 0.47(X^2) - 2.01X + 49.5$$

Figure 10.1 Relationship between household energy consumption per capita and average temperature in 2011 by prefecture

Note: Number of observations is 47. The fitting curve is assumed to be a quadratic function. All coefficients are significant.

and fan heaters owned per household may also contribute to the dominant role of space heating and cooling in residential energy consumption.[2] While in 1990 the average household possessed 1.27 air conditioners and 0.75 fan heaters, the number of air conditioners and fan heaters owned per household increased to 2.68 and 1.26 in 2011, respectively (ECCJ 2013).

Consequently, household energy consumption per capita varies widely across prefectures (see Figure 10.1). In Japan, there are 47 prefectures. In the figure, prefectures on the left and right are located in cold and hot regions, respectively. Therefore, household energy consumption in cold and hot regions is larger than that in other regions (however, the importance of cooling is relatively smaller than that for heating). These data suggest that reducing energy consumption for cooling and heating is an important strategy for the Japanese government as it attempts to reduce household GHG emissions.

To reduce the household energy consumption related to cooling and heating, one useful measure may be to promote the construction of eco-friendly houses. In Japan, 71 percent and 13 percent of heat influx into a house in summer is via windows and walls, respectively. Heat radiation loss through windows and walls is 48 percent and 19 percent in winter, respectively (Vinyl Environmental Council 2013). Therefore, eco-friendly houses must improve heat influx and radiation from windows and walls. Many studies have examined the relationship between heat influx/radiation and characteristics of walls/windows (e.g. Akbari and Taha 1992; Sekhar and Toon 1998; Kossecka and Kosny 2002).

Construction materials such as insulation, double sashes, and double glazing are expected to reduce energy consumption by improving household heat influx and radiation. These materials are representative components of an eco-friendly house. Insulation is a housing/building material used inside walls. Because insulation is composed of material with low thermal conductivity, it can suppress loss of cold air in summer and warm air in winter through walls. Additionally, double-sash windows are designed with the window frame in two pieces.[3] Double-glazing consists of using two sheets of glass for windows with a vacuum space between. Both double-sash and double-glazed windows are better insulated than conventional windows. In fact, installation of these types of windows is essentially required to participate in the subsidy program, which will be described further in Section 10.2.

For example, Howden-Chapman et al. (2007) show that a house with insulation consumes 19 percent less energy than one without insulation, based on a field experiment in New Zealand.[4] Natural Resource Canada (2011) reports that energy-efficient windows (i.e. double-hung/sash windows with double glazing) can reduce a household's energy expenditure for heating by up to 50 percent, a drastic reduction in energy usage.

The installation cost of these materials is higher compared to conventional building materials, such as walls without insulation and single-sash and single-glaze windows. We focus on insulation an example. The cost of insulation (ignoring construction cost) greatly depends on the materials used. For example, the 40 mm thick raw glass wool produced by Asahi Fiber Glass Co. Ltd., Product Code GW32, costs approximately 1,244 yen (approximately $12.40) per square meter, whereas the 40 mm thick phenolic foam produced by Sekisui Chemical Co. Ltd., Product Code JJ40MW, costs approximately 2,930 yen (approximately $29.30) per square meter.[5] Of course, the thermal insulation performance of phenolic foam is better (more energy efficient) than glass wool. In the case of a representative Japanese detached house with a wall area of 150 square meters, the homeowner would spend approximately an additional 186,600 yen and 439,500 yen ($1,866 and $4,395) to install glass wool and phenolic foam, respectively, relative to a homeowner who lives in a house without such insulation.

Construction of an eco-friendly house with the materials described above is more costly compared to construction of a conventional house. Between 2009 and 2011, the Japanese government, therefore, subsidized new construction of eco-friendly houses, with a goal of reducing GHG emissions from the residential sector. Approximately 291 billion yen ($2.91 billion) were provided for the subsidy.

We must emphasize that the subsidy could not always promote construction of eco-friendly housing for two reasons. First, some households would construct new eco-friendly houses even if there was no public financial support. Some of these households, however, would willingly receive the subsidy if it were available. In this case, the subsidy to the households that would have constructed an eco-friendly house anyway did not contribute to the construction of an additional eco-friendly house, suggesting that some portion of the subsidy is composed of a social sunk cost. Second, the amount of the subsidy per household might have been too small to affect a household's decision to order a new eco-friendly house. The amount of subsidy per household was only 300,000 yen ($3,000). According to a report from the Japan Institute of Life Insurance (2014), the average cost to purchase a new home is approximately 35 million yen ($291,700). Many people purchase only one house throughout their lives. Therefore, the subsidy rate relative to the cost of a new home was less than 1 percent, which was less than the subsidy rates for other rebate programs for home appliances (see Part II) and automobiles (see Part III). The subsidy rates for these two programs range up to 10 percent.

There is little evidence of whether the subsidy indeed encouraged construction of eco-friendly houses. Mizuho Research Institute (2012) showed that the subsidy reduced GHG emissions from households by only 0.004 percent, implying that the reduction was negligible. However, the authors did not consider other factors affecting the number of eco-friendly houses, such as income and economic fluctuations, because they used aggregate national data. Using Japanese prefectural panel data, this chapter, therefore, is the first study to empirically examine the effects of the subsidy while controlling for other factors.[6]

The remainder of this chapter consists of the following sections. Section 10.2 describes the subsidy for eco-friendly houses implemented in Japan. Section 10.3 provides our empirical estimation model for analyzing whether the subsidy increased the number of eco-friendly houses built in Japan. Sections 10.4 and 10.5 present the estimation results and concluding remarks, respectively.

10.2. The subsidy for eco-friendly houses

In this section, we explain in detail the rebate program for eco-friendly houses. To mitigate residential GHG emissions, the Japanese government implemented an eco-point program for housing in 2009 (Housing Eco-point Office 2014).[7] The program ended in 2011. The goal of the program was to promote the construction of energy-efficient, environmentally-friendly houses. Under the program, a homeowner constructing a new house that satisfied certain criteria designated by the program could receive a rebate worth 300,000 yen ($3,000) by submitting an application to the government. If a solar photovoltaic system was installed in a new house, the amount of the subsidy increased to 320,000 yen ($3,200). Although there is room for debate as to whether the amount of subsidy is enough compared to the cost of a new home, approximately 291 billion yen ($2.91 billion) was budgeted for the subsidy program. Thus, the significant amount of funding set aside demands an examination of the effect of the rebate program.

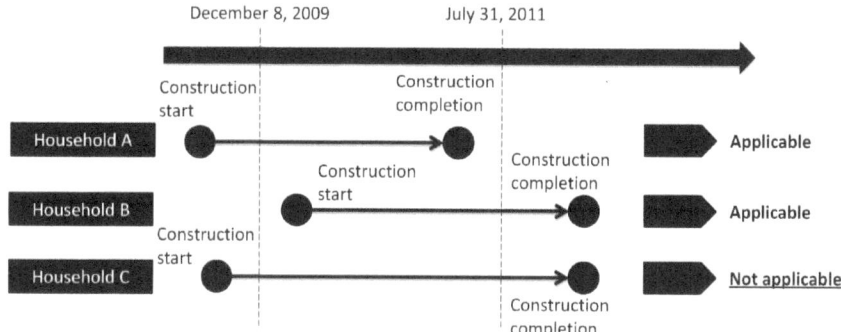

Figure 10.2 Who can apply to the program?

The government subsidized households that constructed new eco-friendly houses from 8 December 2009 to 31 July 31 2011 (applications were accepted through 31 March 2014). Households that began construction during that period were eligible. Figure 10.2 provides an overview of program eligibility. Household A and B in the figure could apply. Even if household C constructed an eco-friendly house, he or she could not receive the subsidy because of the construction period.[8]

To qualify for the subsidy program, a new house had to satisfy the strictest housing performance standards established by the Energy Conservation Act and the Housing Quality Assurance Act. These acts created four grades for home energy performance (i.e. Grade 1 for lowest performance to Grade 4 for highest performance).[9] Grade 4, the most stringent standard, applied to houses with better energy performance than required by the 1999 building standards. Houses labeled Grades 2 and 3 performed at least as well as required by the 1980 and 1992 standards, respectively. Grade 1 designated a house that performed worse than the 1980 standards. To be eligible for the subsidy program, a house had to meet the standard for Grade 4. The factors that were most important to qualify for a particular grade varied according to the region in which a house was located.[10] For example, energy performance for heating was important for cold regions (i.e. Region 1 and 2 in Figure 10.3), while energy performance for cooling was a key determinant in hot regions (i.e. Region 5 and 6 in Figure 10.3). This is similar to the Energy Star Certified Homes program in the United States (Energy Star 2014). The regional classifications are presented in Figure 10.3. Japan's 47 prefectures were aggregated into six classifications.

A new house for which the owner intended to apply for the subsidy had to be certified as a Grade 4 eco-friendly house with excellent energy performance. To satisfy the requirements of Grade 4, the house had to be built with high-quality insulation and double-glazed and double-sash windows (this is essential in cold regions), as described above. Therefore, the Japanese government and builders recommended installation of these materials as well as energy-efficient equipment for new homes that were being constructed.

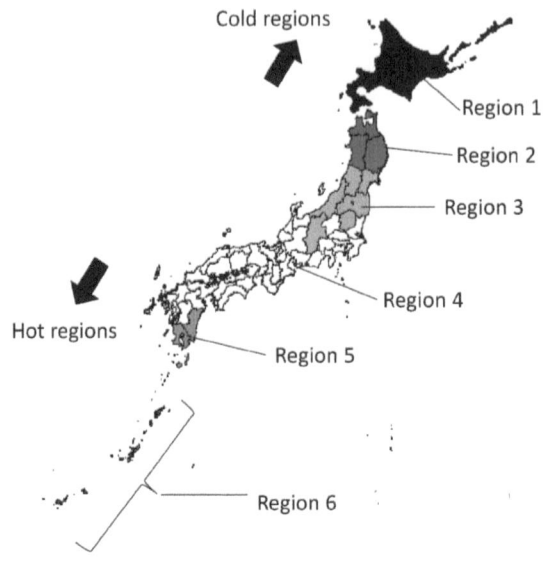

Figure 10.3 Regional classifications

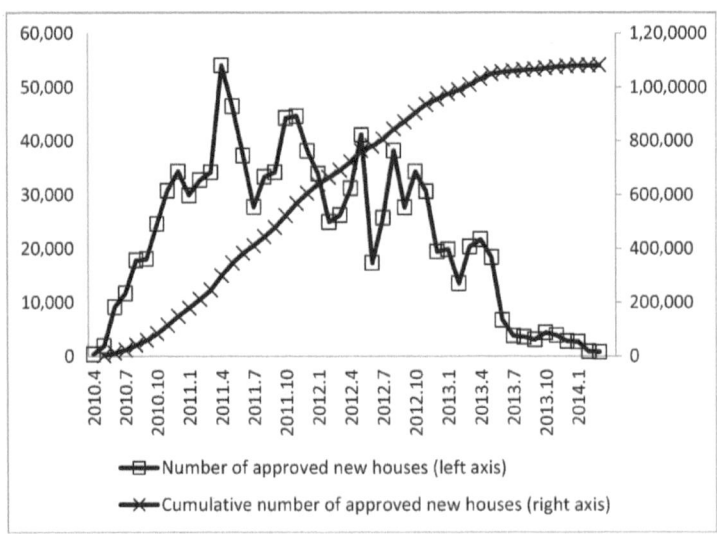

Figure 10.4 Number of approved new eco-friendly houses by month

At the end of July 2014, the amount of subsidy for new eco-friendly houses cumulatively reached approximately 291 billion yen ($2.91 billion). The total number of approved new eco-friendly houses was 1,085,732. The number of approved new eco-friendly houses by month is shown in Figure 10.4, illustrating that many new eco-friendly houses were approved after 2011. There was a time

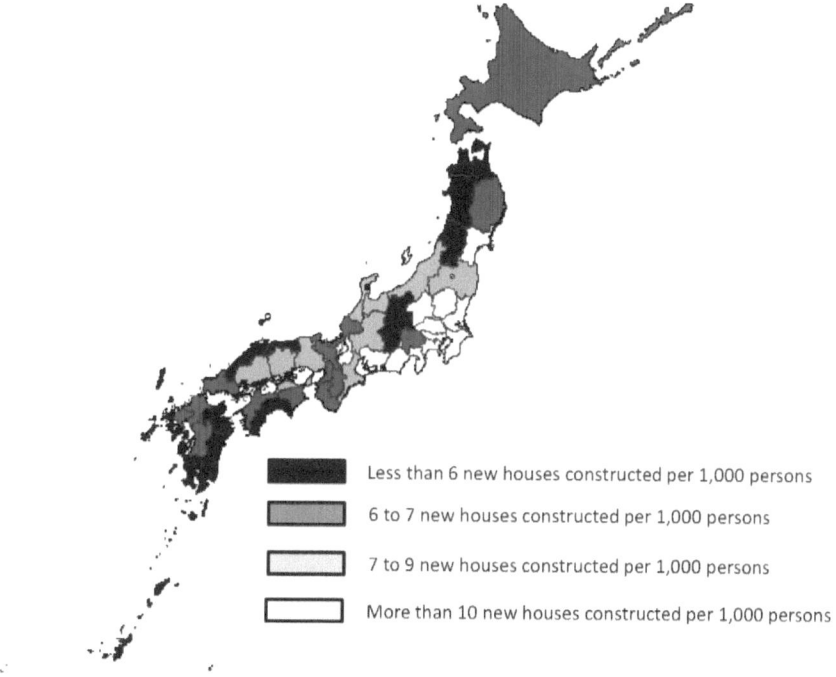

Less than 6 new houses constructed per 1,000 persons

6 to 7 new houses constructed per 1,000 persons

7 to 9 new houses constructed per 1,000 persons

More than 10 new houses constructed per 1,000 persons

Figure 10.5 Number of approved new eco-friendly houses per 1,000 persons by prefecture

lag between the applicants' submission of the application for the subsidy, timing of construction completion (see Figure 10.2), and government approval. Therefore, although the subsidy program began in December 2009, the first approved eco-friendly houses were counted in April 2010. Similarly, while the program ended in July 2011, houses were still being approved as late as April 2014. It is also possible that some owners did not apply for the subsidy even though their new houses were eco-friendly. Any such houses that were not submitted are not included in the figure.

Figure 10.5 shows, by prefecture, the cumulative number of approved newly constructed eco-friendly houses per 1,000 persons. Household energy consumption is expected to be higher in cold and hot regions because space heating and cooling use large amounts of energy. However, households in the most northern and southern regions do not appear more likely to construct eco-friendly houses than households in more temperate regions. Instead, we find a large number of new eco-friendly houses constructed in the region near Tokyo. The high incomes of residents near Tokyo may explain this trend. Therefore, it may be important to take other factors, such as income, into account when examining the relationship between the subsidy and the number of newly constructed eco-friendly houses.

10.3. Empirical analysis

10.3.1. Estimation model

In this section, the empirical estimation equations for examining whether the subsidy program promoted construction of eco-friendly houses are introduced. These equations were used to regress the program on the number of new eco-friendly houses to determine its effect. We estimate two equations, where the dependent variables are the number of eco-friendly houses (HE_{it}) and conventional non-eco-friendly houses (HO_{it}) in a given prefecture in a given year. The subscripts i and t in each variable stand for prefecture and year, respectively. The dummy variable for the subsidy program is SP_{it} (which takes a value of 1 if the program is in effect). To account for potential differences in the effect of the subsidy among regions, interaction terms of six regional dummies ($RD_{k,i}$ [$k = 1,...,6$]) and a subsidy dummy (SP_i) are included in the model. Region refers to the six classifications presented in Section 10.2. If a prefecture belongs to Region 1, $RD_{1,i}$ takes a value of 1, otherwise it takes a value of 0. The other control variable vector is written as OT_{it} and is described below. Additionally, we use T_t to capture economic fluctuations. Unobserved prefectural specific effects and idiosyncratic error are expressed as μ_i and ε_{it}, respectively.

A linear relationship is assumed between the number of new eco-friendly houses and the determinants, including the dummy variable for subsidy availability. The estimation equation can be written as follows:

$$HE_{it} = \alpha_1 + \alpha_2 OT_{it} + \sum_{k=1}^{6} \beta_k RD_{k,i} \times SP_{it} + \delta T_t + \mu_i + \varepsilon_{it},$$

where α_k (k = 1, 2), β_l ($l = 1,...,6$), and δ are parameters. When using the number of conventional houses as a dependent variable, the left-hand-side variable is HO_{it} instead of HE_{it}. We estimate the equations using ordinary least squares. In addition to the number of houses (HE_{it} and HO_{it}), the area (in square meters) of the houses (HE_A_{it} and HO_A_{it} for eco-friendly and conventional houses, respectively) are employed as dependent variables because it is possible that the size of eco-friendly houses was influenced by the subsidy even if the subsidy did not affect the number of eco-friendly houses. Furthermore, we use the proportion of eco-friendly house to total houses (RHE_{it}) as a dependent variable (proportion of conventional houses can be written as $1 - RHE_{it}$). Therefore, five equations are estimated in this chapter.

The estimated sign and significance of the coefficient for the subsidy dummy (β_k) is key to evaluating the subsidy. If we observe that the estimated coefficient is positive and significant, we can conclude that the subsidy increased the number of eco-friendly houses. If not, we must conclude that the subsidy did not have its intended effect. The estimated coefficients also allow us to verify any differences in the effect of the subsidy among regions, the possibility of which was illustrated in Figure 10.5. If the effect of the subsidy on new eco-friendly houses in region k is different than that in region j ($k \neq j$), the estimated coefficients will not be equal; that is, $\widehat{\beta_k} \neq \widehat{\beta_j}$. Therefore, we can evaluate the effect of the subsidy program by examining the coefficients of parameter β_k.

We also control for population, population density, income per capita, and road length. The population variable also controls for the size of the prefecture. It is expected that the number of houses constructed will be higher for prefectures with high populations versus those with low populations. However, it is not clear whether the same should be expected for eco-friendly houses. Population density, or population divided by residential area, captures any difference in new home construction between urban and rural prefectures.

Many studies have shown that high-income households are more likely to purchase rather than rent houses (Goodman 1988).[11] This trend has been observed in Japan (Horioka 1988; Seko and Sumita 2007). As noted above, eco-friendly houses cost more than conventional ones. Therefore, income may similarly influence the decision of whether to construct a new eco-friendly house.

We use main road length as a proxy for household mobility (e.g. accessibility and opportunity). Suppose a household works at a prefecture's central business distinct and is seeking a place to live. Long road length in the prefecture indicates that the household may have the option to live in their own detached house constructed outside of the central business distinct because they can easily commute to their workplace by automobile. Therefore, the number of eco-friendly houses may be larger in prefectures with long road lengths compared to those with short road lengths.[12]

10.3.2. Descriptive statistics

We use prefectural balanced panel data from 2000 to 2010. Therefore, the sample size is 517 (= 47 prefectures × 11 years). The descriptive statistics are shown in Table 10.1. The number of newly constructed houses can be further categorized into the number of new eco-friendly houses and the number of new conventional houses. Similarly, the area of newly constructed houses can be further categorized as the area of eco-friendly houses and the area of conventional houses. As noted above, these five variables (i.e. *HE*, *HO*, *HE_A*, *HO_A*, and *RHE*) are the dependent variables in our estimations.

The data for these five dependent variables are not directly obtained from statistical data. Data on the number and area of newly constructed houses (i.e. *HA* and *HA_A*) are derived from the *Survey on Construction Statistics* provided by the Ministry of Land, Infrastructure, Transportation and Tourism (MLIT). Additionally, MLIT (2012) reports the annual average ratio of new eco-friendly houses in Japan from 1999 to 2010. Because the ratio reflects the average for Japan as a whole, we reduce it to the prefectural level using the number of approved eco-friendly houses in Figure 10.5. Then, applying the calculated ratio for each prefecture, the number of newly constructed houses is divided into two variables, *HE* and *HO*. The same procedure is applied for the area of new eco-friendly and conventional houses.

Table 10.1 Summary statistics

Variable	Description	Unit	Mean	S.D.	Min
HA	Number of new constructed houses	1,000 house	23.41	29.95	2.06
HE	Number of eco-friendly houses	1,000 house	4.20	7.35	0.11
HO	Number of conventional houses	1,000 house	19.20	24.39	1.59
HA_A	Area of new constructed houses	1,000 square meter	2,073.25	2,342.74	199.52
HE_A	Area of eco-friendly houses	1,000 square meter	364.26	572.79	10.36
HO_A	Area of conventional houses	1,000 square meter	1,708.99	1,933.88	153.52
RHE	Proportion of eco-friendly houses	percent	0.16	0.11	0.01
POP	Population	1,000 person	2,717.77	2,575.75	588.67
DEN	Population density	person/hectare	237.23	114.46	124.55
INC	Income per capita	yen/person	2.73	0.41	1.99
ROD	Road length	kilometer	3,899.37	2,433.89	1,450.50
CI	Composite index for economic activities	2010 = 100	103.93	8.27	85.61

Note: Number of the observations is 517.

The population data are obtained from *Population Estimates* provided by the Statistics Bureau of the Ministry of Internal Affairs and Communications. *Prefectural Accounts* and the *Annual Report on Urban Planning* provide data on income and residential area for each prefecture. Road length is obtained from the *Survey on Current Conditions of Road Facilities.* Roads are roughly classified into four types: expressway, national way, prefectural road, and city road. For road length, we use the total length of national roads and prefectural roads within each prefecture. Data on the composite index for economic activities, which captures economic fluctuations, are obtained from *Business Statistics,* provided by Cabinet Office in Japan.

10.4. Empirical analysis

10.4.1. Estimation results

The estimation results are shown in Table 10.2. Columns (1) and (2) show the results where the dependent variables are the number of eco-friendly houses and conventional houses, respectively. In Column (1), the coefficient on the interaction of the subsidy dummy and regional dummy for Region 1 is approximately 6.2 and is significant at the 1 percent level, implying that the subsidy increased the number of new eco-friendly houses constructed in Region 1 by approximately 6,200. We also found that the subsidy increased the number of eco-friendly houses constructed in Regions 2, 3, 4, and 5 by approximately 3,500, 4,500, 4,100 and 3,800 houses, respectively. The observed increase in these regions is smaller than in Region 1. In contrast, the subsidy program had a non-significant effect on the number of eco-friendly houses constructed in Region 6, suggesting that the subsidy did not influence the households' decision on eco-friendly house constructions in Region 6. Therefore, the subsidy promoted eco-friendly house construction in colder regions but did not increase eco-friendly house construction in hotter regions.

The interaction coefficients in Column (2), except Region 6, are found to be significant. In Region 1, where eco-friendly house construction increased with the subsidy, the number of conventional houses decreased by 26,500 houses after the implementation of the subsidy. In Regions 2 through 5, conventional house construction also slowed with implementation of the subsidy. In contrast, the coefficient of the interaction term for Region 6 is not significant, implying that conventional house construction did not change with the subsidy. These results in Region 1 through 5 are opposite to those in Column (1). This result may reflect a substitution of conventional houses for eco-friendly houses because many households, in general, purchase only one house over their lifetime. That is, an increase in eco-friendly house construction may lead to a decrease in conventional house construction through substitution effects. Therefore, the subsidy decreased conventional house construction in cold regions and did not affect conventional house construction in hotter regions.

Table 10.2 Estimation results

Variables	Number of Houses		Area of House		Rate of Eco.
	(1) Eco HE	(2) Conv. HO	(3) Eco HE_A	(4) Conv. HO_A	(5) RHE
SP × Region 1	6.196***	−26.545***	504.642***	−2,167.719***	0.223***
	(1.415)	(3.961)	(124.261)	(312.897)	(0.024)
SP × Region 2	3.538***	−8.742***	288.387***	−808.737***	0.224***
	(0.641)	(1.295)	(52.672)	(103.804)	(0.012)
SP × Region 3	4.535***	−10.220***	401.789***	−1,000.509***	0.299***
	(0.489)	(0.987)	(41.812)	(78.051)	(0.023)
SP × Region 4	4.072***	−10.551***	377.625***	−903.036***	0.264***
	(0.519)	(1.112)	(47.396)	(86.284)	(0.010)
SP × Region 5	3.810***	−8.230***	316.670***	−782.630***	0.226***
	(0.623)	(1.589)	(51.341)	(143.028)	(0.011)
SP × Region 6	0.662	−0.890	71.449	−64.261	0.176***
	(0.925)	(1.497)	(78.941)	(161.713)	(0.015)
POP	0.035	−0.073***	2.427***	−6.606***	0.0002***
	(0.006)	(0.010)	(0.452)	(0.534)	(4.07e−05)
DEN	0.023	−0.074*	2.857**	−4.725	0.001
	(0.015)	(0.039)	(1.401)	(3.034)	(0.0003)

	(1)	(2)	(3)	(4)	(5)
INC	-1.950	9.080***	-216.603	943.256***	-0.148***
	(1.848)	(3.443)	(157.738)	(217.553)	(0.024)
ROD	0.017***	-0.028**	1.437***	-4.133***	0.0003***
	(0.004)	(0.012)	(0.359)	(1.050)	(8.57e-05)
CI	0.134***	-0.121***	11.720***	-15.140***	0.006***
	(0.015)	(0.033)	(1.331)	(2.851)	(0.0003)
CONSTANT	-342.860***	734.022***	-25,091.810***	75,009.650***	-2.682***
	(42.361)	(93.756)	(3,506.940)	(6,582.174)	(0.537)
F-value	51.52***	170.19***	46.01***	152.68***	
Adj. R-squared	0.927	0.968	0.911	0.970	
Log pseudo likelihood					992.4

Note: Number of the observations is 517. Robust standard errors are in parentheses. ***, **, and * represent 1%, 5%, and 10% significance levels. Prefectural fixed effects are included in all models.

As for the other control variables, the sign of population in Column (1) is positive at the 1 percent significance level. The coefficient of road length is also significant and positive, whereas it is found that population density and income do not influence the number of eco-friendly houses constructed. Therefore, the number of new eco-friendly houses in prefectures with high population and long road length – that is, urban prefectures – is larger than in rural prefectures. Further, the coefficient on the composite index is positive at the 1 percent significance level, suggesting that construction of eco-friendly houses is likely to increase with the booming economy.

The estimated signs of most of the control variables in Column (2) are inverse to those of eco-friendly houses. That is, in rural prefectures with low population/population density and short road length, construction of conventional houses was high, relative to urban prefectures. In contrast to the results in Column (1), the coefficient of income is found to be significant and positive at the 1 percent level, suggesting that households with high income are likely to construct conventional houses. The composite index is negatively related to conventional house construction, which is also the opposite of the result in Column (1). Therefore, the construction of conventional houses is decreasing with the booming economy.

Estimation results using the areas rather than numbers of eco-friendly and conventional houses as the dependent variables are presented in Columns (3) and (4), respectively. The overall trend of the estimation results in Columns (3) and (4) is similar to those in Columns (1) and (2), respectively. The subsidy increased the total area of eco-friendly houses in Regions 1 through 5. At the same time, the area of conventional houses decreased in Regions 1 through 5. In Region 6, both eco-friendly and conventional house constructions were not influenced by the subsidy. Therefore, even when evaluating the effect of the subsidy program from the viewpoint of home size, it is found that the subsidy promoted eco-friendly houses while it decreased construction of conventional houses, except in hotter regions.

The share of eco-friendly houses to total houses ranges from 0 to 1. When we use the share as a dependent variable, the ordinary least square is not an appropriate estimation method due to the characteristics of this statistical method. Thus, instead of ordinary least square, we employ generalized linear models where the distribution of the dependent variable is assumed to be Gaussian (Papke and Wooldridge 1996). For computation, the glm command in the STATA software package is used.

The estimation results using share of eco-friendly houses to total houses as a dependent variable is presented in Column (5). The estimated signs are similar to those of Columns (1) and (3). The coefficient on the interaction of the subsidy dummy and regional dummy for Region 1 is approximately 0.223 and significant at the 1 percent level, implying that the subsidy increased the share of eco-friendly houses in Region 1 by 22.3 percent. In contrast to the results shown in Columns (1) and (3), it is found that the subsidy increased the share of eco-friendly houses in Region 6 by 17.6 percent. Therefore, in evaluating

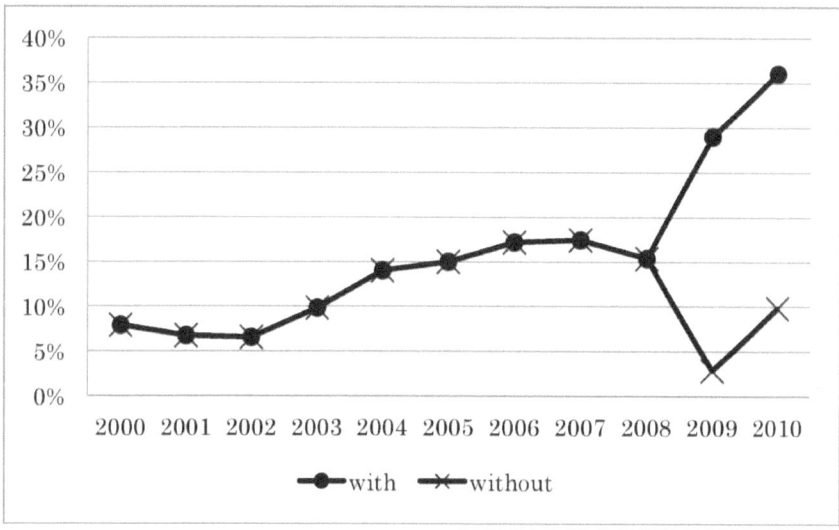

Figure 10.6 Predicted proportions of eco-friendly houses with and without the subsidy

the effect of the subsidy from the viewpoint of proportion, it is revealed that the subsidy increased the share of eco-friendly houses to total houses anywhere in Japan.

Using the estimation results in Column (5) of Table 10.2, Figure 10.6 shows the transitions of the predicted proportion of eco-friendly houses, on average. The lines labeled as "with" and "without" represent the predicted proportions with and without the subsidy, respectively. Therefore, the difference between the two lines indicates the effects of the subsidy. The proportion decreased after 2009 without the subsidy, which may have been caused by the worldwide recession known as the Lehman shock. The proportion rose after the implementation of the subsidy. Consequently, the subsidy increased the proportion by 26 percent in 2009 to 2010, on average.

10.4.2. Cost-effectiveness of the subsidy

We found that the implementation of the subsidy promoted construction of eco-friendly houses in cold regions and decreased construction of conventional ones. Also, we found that the number of eco-friendly and conventional houses built during the same period did not change in hotter regions. Therefore, both the increases and decreases in eco-friendly and conventional houses must be taken into account to determine the effect of the subsidy program on residential GHG emissions.

Table 10.3 shows the predicted numbers of both house types by region. The top and bottom tables represent the predicted numbers of eco-friendly (*HE*)

Table 10.3 Estimated regional change in GHG emissions with the subsidy from 2009 to 2010

	HE with	HE without (thousand houses)	Change	GHG Emissions Change (thousand CO_2-ton)
Region 1	68.9	56.5	12.4	7.1
Region 2	30.5	9.3	21.2	12.2
Region 3	196.5	142.1	54.4	31.4
Region 4	2,017.6	1,740.7	276.9	159.5
Region 5	26.2	11.0	15.2	8.8
Region 6	0.8	-0.6	1.3	0.8
Total	2,340.6	1,959.1	381.5	219.8
	HO with	HO without (thousand houses)	Change	GHG Emissions Change (thousand CO_2-ton)
Region 1	612.5	665.6	–53.1	–76.5
Region 2	313.8	366.2	–52.5	–75.6
Region 3	1,153.7	1,276.3	–122.6	–176.7
Region 4	12,282.5	13,000.0	–717.5	–1,033.5
Region 5	264.7	297.6	–32.9	–47.4
Region 6	207.8	209.6	–1.8	–2.6
Total	14,835.0	15,815.4	–980.3	–1,412.1

Note: Note that "Change" is the difference between "*HE* with" and "*HE* without." If a reduction in GHG emissions was obtained, the values of the far right column are less than 0.

and conventional houses (*HO*), respectively. The columns labeled as "*HE* with" and "*HE* without" represent the predicted numbers of eco-friendly houses with and without the subsidy program, using the estimation results in Column (1) of Table 10.2. The subsidy increased the number of eco-friendly houses built from 2009 to 2010 by 12,400 houses in Region 1. The subsidy increased the number of eco-friendly houses built by 381,500 houses overall.

The Japanese government approved more than 1 million eco-friendly houses to receive subsidies (see Figure 10.4), whereas the estimated total increase in eco-friendly houses as a result of the subsidy program was only 381,500. Therefore, regardless of the presence or absence of the subsidy program, there might have been many households that intended to construct eco-friendly houses (roughly about 65 percent of households), suggesting that we should not evaluate the subsidy program by viewing only the approved number of the eco-friendly houses. To examine the effects of the program, it is important to control for other factors, such as population and time trends.

Although eco-friendly houses have better energy performance than conventional houses, the increase in eco-friendly houses caused an increase in GHG emissions. According to MOE (2006), the energy performance of a representative eco-friendly house and conventional house is 220 and 550 liters (crude oil equivalent), respectively. Although performance varies by region, we use these representative values to calculate the effect of the subsidy program on GHG emissions. We find that new construction of eco-friendly houses generated an increase of 219,800 tons of CO_2 in two years. When we exclude an increase of GHG emissions in Region 6, where the significant effects of the subsidy is not found, the GHG emissions increase is 219,100 tons of CO_2.

At the same time, the number of conventional houses decreased with implementation of the subsidy program. This decrease may reflect replacement of conventional houses with eco-friendly ones. Similar to the eco-friendly houses, the columns labeled as "*HO* with" and "*HO* without" stand for the predicted numbers with and without the subsidy program, which is predicted by the estimation results at Column (2) of Table 10.2. The subsidy decreased construction of conventional houses in Japan by 980,300, leading to a decrease in residential GHG emissions of 1,412,100 tons of CO_2 over two years. Overall, the subsidy program reduced residential GHG emissions by 1,192,300 CO_2-tons (= 1,412,100 – 219,800). The subsidy program reduced, on average, 596,200 tons of CO_2 per year.

In Japanese accounting, the legal durable lifetime of common synthetic resin or wooden houses is 22 years. Therefore, in 22 years 13,115,500 tons of CO_2 (= 596,200 × 22) emissions will be avoided by the subsidy program. Because the program cost 291 billion yen ($2.91 billion), the marginal abatement cost of GHG mitigation is 22,188 yen ($221.90) per ton of CO_2.[13] When the effects in Region 6 are omitted, the marginal abatement cost is 22,221 yen ($222.20) per ton of CO_2.

Economic theory dictates that the subsidy program improved social welfare only if its marginal abatement cost was smaller than the marginal externality cost of GHG emissions. Previous studies have estimated a wide range of marginal externality costs for GHG emissions (Tol 2009). Tol (2009) shows that the mean social cost of carbon estimated by 232 previous studies is approximately 10,500 yen ($105), implying that the marginal abatement cost of the subsidy program estimated in this chapter may be larger than the marginal externality cost. Therefore, the subsidy program may not be a cost-effective measure to reduce residential GHG emissions. However, this calculation greatly depends on the assumption of the life expectancy of a house (i.e. 22 years). Therefore, if actual life expectancy is longer than our assumption, the estimated marginal abatement cost would be smaller, and the cost-effectiveness would be improved. This implies that a complementary policy to extend the expected lifetime of houses may be important to cost-effectively reducing residential GHG emissions.

10.5. Conclusions

In this chapter we examined the effect of a subsidy program for eco-friendly houses, which the Japanese government implemented in 2009 to reduce residential GHG emissions. The program's total budget was large, 291 billion yen ($2.91 billion). Despite the program's large budget, there is little evidence of the extent to which the subsidy promoted eco-friendly houses and reduced residential GHG emissions. It is important to examine both of these factors to determine whether the program's benefits were commensurate with its costs.

Using Japanese prefectural level data from 2000 to 2010, we performed an econometric analysis using a regression model to examine the cost-effectiveness of the subsidy program. We found that the subsidy program increased the number of eco-friendly houses, on average. Although the energy performance of eco-friendly houses is better than that of conventional houses, GHG emissions from eco-friendly houses as a whole increased because the program increased the total number of eco-friendly houses in Japan. In addition, the effect of the subsidy on construction of eco-friendly houses varied among regions. In particular, the number of eco-friendly houses constructed largely increased in cold regions.

In contrast, we found that the number of non-eco-friendly conventional houses constructed was likely to decrease after the conclusion of the program, which might reflect a substitution effect because conventional houses are substitutes for eco-friendly houses. Due to the decrease in the number of conventional homes constructed, GHG emissions from conventional houses also decreased. In contrast to the change in eco-friendly houses, the number of the conventional house constructions greatly decreased in cold regions.

Summing the increase and decrease of GHG emissions from new eco-friendly and conventional houses, we confirmed that the subsidy program reduced overall residential GHG emissions by 596,000 tons of CO_2 per year. Based on the assumption that new houses have a lifetime of 22 years, the marginal abatement cost of the subsidy (i.e. cost per ton of GHG emissions reduction) is approximately 22,000 yen ($220). This is higher than the social cost of carbon, approximately 10,000 yen ($100), estimated by Tol (2009). Therefore, the subsidy program may be not a cost-effective policy to address global warming. However, extending the life expectancy of new houses could improve the program's cost-effectiveness.

The limitations of the analysis described in this chapter are twofold. First, we do not explicitly consider household decision-making with regard to whether to construct an eco-friendly or a conventional house. Nor do our models account for the tenure-choice problem (Henderson and Ioannides 1983), which relates to whether households choose to purchase or rent. Tackling these limitations requires household-level rather than aggregate data. Second, more detailed geographical conditions such as temperature, humidity, and precipitation would be useful in examining the effect of the subsidy and to derive more practical policy lessons related to the promotion of eco-friendly houses. Although our economic models incorporate prefectural fixed effects, these omitted geographical conditions are conflated with the fixed effects in our models. Clarifying for policymakers appropriate subsidies based on geographical factors could aid in the design of more cost-effective programs.

Notes

1 Two additional characteristics (i.e. ecological lifestyle and harmony with local community) are presented in MOE (2014). However, we omit these characteristics in this chapter.
2 Fan heaters include oil, gas, and electric heaters.
3 Double-hung windows are also composed of two window sashes; in this design one window is on the top and another is on the bottom. This window may be more popular in the Europe and United States; double-sash windows are used in Japan.
4 Appelfeld and Svendsen (2011) found, based on a laboratory experiment, that ventilated windows can reduce energy consumption by approximately 10 percent. Galvin (2013) suggests that the number of external surfaces may be relevant to household consumption.
5 100 yen is converted into $1.
6 Although many studies have examined the relationship between subsidies and housing decisions (e.g. White and White 1977; Rosen 1979), these studies mainly focused on the role of subsidies for low-income households.
7 For the details, refer to Housing Eco-Point Office (2014).
8 Households that transformed their existing conventional houses into eco-friendly ones were also eligible for applying to the subsidy program; however, we do not consider these renovated homes due to the lack of a sufficient available dataset.
9 The Housing Quality Assurance Act also establishes standards with four scales for nine home characteristics: elder care, earthquake resistant (collapse prevention), earthquake resistant (damage prevention), wind resistant, refractory (warning device), refractory (aperture), degradation measure, maintenance measure, and formaldehyde measure.
10 For details, refer to the Institute for Building Environment and Energy Conservation (2014).
11 This is called the "tenure-choice problem" in urban and housing economics (see Henderson and Ioannides 1983).
12 This is a kind of residential choice problem (see Reschovsky 1979).
13 For simplicity, the discount rate is not considered here.

References

H. Akbari and H. Taha (1992). 'The impact of trees and white surfaces on residential heating and cooling energy use in four Canadian cities', *Energy* 17, 141–9.
D. Appelfeld and S. Svendsen (2011). 'Experimental analysis of energy performance of a ventilated window for heat recovery under controlled conditions', *Energy and Buildings* 43, 3200–7.
M. Costanzo, D. Archer, E. Aronson, and T. Pettigrew (1986). 'Energy conservation behavior: the difficult path from information to action', *American Psychologist* 41, 521–38.
Energy Conservation Center Japan (2013). *Handbook of energy and economic statistics in Japan*, Energy Conservation Center Japan, Tokyo (in Japanese).

Energy Star (2014). 'Energy Star certified homes version 3 program requirements', Available: <https://www.energystar.gov/index.cfm?c=bldrs_lenders_raters.nh_v3_guidelines> (accessed 7 October 2014).

R. Galvin (2013). 'Targeting "behavers" rather than behaviours: A "subject-oriented" approach for reducing space heating rebound effects in low energy dwellings', *Energy and Buildings* 67, 596–607.

A.C. Goodman (1988). 'An econometric model of housing price, permanent income, tenure choice, and housing demand', *Journal of Urban Economics* 23, 327–53.

J.V. Henderson and Y.M. Ioannides (1983). 'A model of housing tenure choice', *American Economic Review* 73, 98–113.

C.Y. Horioka (1988). 'Tenure choice and housing demand in Japan', *Journal of Urban Economics* 24, 289–309.

Housing Eco-point Office (2014). 'What is the housing eco-point?' Available: <http://jutaku.eco-points.jp/top.html> (in Japanese) (accessed 7 October 2014).

P. Howden-Chapman, A. Matheson, J. Crane, H. Viggers, M. Cunningham, T. Blakely, C. Cunningham, A. Woodward, K. Saville-Smith, D. O'Dea, M. Kennedy, M. Baker, N. Waipara, R. Chapman, and G. Davie (2007). 'Effect of insulating existing houses on health inequality: cluster randomised study in the community', *BMJ: British Medical Journal* 334, 460–4.

Institute for Building Environment and Energy Conservation (2014). 'Criteria of housing owners', Available: <http://ees.ibec.or.jp/index.php> (in Japanese) (accessed 7 October 2014).

Japan Institute of Life Insurance (2014). 'How much is a residential house?' Available: <http://www.jili.or.jp/lifeplan/lifeevent/house/3.html> (in Japanese) (accessed 7 October 2014).

E. Kossecka and J. Kosny (2002). 'Influence of insulation configuration on heating and cooling loads in a continuously used building', *Energy and Buildings* 34, 321–31.

Ministry of Land, Infrastructure, Transport and Tourism (2012). 'Energy saving measures to houses and buildings', Available: <http://www.mlit.go.jp/common/000193924.pdf> (in Japanese) (accessed 7 October 2014).

Ministry of the Environment (2006). 'Factor analysis of greenhouse gas emissions in Japan, vol.2', Available: <https://www.env.go.jp/council/06earth/y060–38/mat02–2.pdf> (in Japanese) (accessed 7 October 2014).

Ministry of the Environment (2014). 'What's ecohouse', Available: <http://www.env.go.jp/policy/ecohouse/about/index.html> (in Japanese) (accessed 7 October 2014).

Mizuho Research Institute (2012). 'Evaluation on environmental policies related to housing', Available: <http://www.mizuho-ri.co.jp/publication/research/pdf/report/report12–0222.pdf> (in Japanese) (accessed 7 October 2014).

Natural Resource Canada (2011). 'Improving Window Energy Efficiency: Why Should I Worry About My Windows?', Available: <http://www.nrcan.gc.ca/sites/www.nrcan.gc.ca/files/energy/pdf/energystar/IWEE_EN.pdf> (accessed 7 October 2014).

L.E. Papke and J. Wooldridge (1996). 'Econometric methods for fractional response variables with an application to 401(k) plan participation rates', *Journal of Applied Econometrics* 11, 619–32.

A. Reschovsky (1979). 'Residential choice and the local public sector: An alternative test of Tiebout Hypothesis', *Journal of Urban Economics* 6, 501–20.

H.S. Rosen (1979). 'Housing decisions and the U.S. income tax: An econometric analysis', *Journal of Public Economics* 11, 1–23.

S.C. Sekhar and K.L.C. Toon (1998). 'On the study of energy performance and life cycle cost of smart window', *Energy and Buildings* 28, 307–16.

M. Seko and K. Sumita (2007). 'Japanese housing tenure choice and welfare implications after the revision of the tenant protection law', *Journal of Real Estate Finance and Economics* 35, 357–83.

R.S.J. Tol (2009). 'The economic effects of climate change', *Journal of Economic Perspectives* 23, 29–51.

Vinyl Environmental Council (2013). 'Comfortable life from the resin window', Available: <http://www.jmado.jp/contact/pdf/k-life.pdf> (in Japanese) (accessed 7 October 2014).

M.J. White and L.J. White (1977). 'The tax subsidy to owner-occupied housing: Who benefits?', *Journal of Public Economics* 7, 111–26.

11 Environmental subsidies and life cycle assessment

Tomohiro Tasaki

11.1. Introduction

Environmental consumer subsidy programs subsidize consumers' pro-environmental behaviors. Such programs need to ascertain (1) whether a certain behavior occurred and (2) whether the behavior was pro-environmental in nature. Previous chapters in this volume have discussed the rationales and effectiveness of subsidy programs that provide economic incentives to consumers. In this chapter, we will explore these two factors with respect to policy design. We also address the provision of information to consumers concerning the environmental impacts of the respective behavior.

11.2. Typology of pro-environmental behaviors and the behaviors targeted by consumer subsidy programs

Consumer pro-environmental behaviors are diverse. Tables 11.1 and 11.2 list examples of pro-environmental behaviors. Table 11.1 lists pro-environmental behaviors from two Japanese public opinion surveys addressing environmental issues and energy. These include pro-environmental behaviors with respect to energy use; behaviors with respect to reduction, reuse, and recycling of waste (i.e. the 3Rs); and behaviors concerning ecosystems and biodiversity. The pro-environmental behaviors with respect to energy include refraining from the use of energy-consuming products, minimizing usage, reducing energy consumption through efficient use, and replacing existing products with alternative, energy-efficient products. The pro-environmental behaviors with respect to the 3Rs include avoiding the use of products, using products for an extended period, sharing products, achieving cooperation in source separation of recyclables, and purchasing reused/recycled products. The pro-environmental behaviors with respect to ecosystems include interacting with nature and selecting and consuming seasonal and local produce.

Table 11.2 lists 57 pro-environmental behaviors provided by Aoki et al. (2012). Nineteen of these are related to product consumption, such as the purchase of energy-efficient products and solar panels. Fourteen behaviors involve the use of products, including the adjustment of air conditioning temperature controls and

Table 11.1 Examples of pro-environmental behaviors (retrieved and translated from public opinion surveys on energy [2005] and environmental issues [2014] conducted by the Japanese Cabinet Office)

Energy

Reducing time spent watching television and listening to the radio, and then diligently turning them off.

Refraining from using the lights and air conditioners, and in case of use, diligently turning them off.

Only turning on the electricity for the part of the electric carpet that needs it.

Finishing bathing quickly so that all members of the family can take a bath one after another.

Correctly positioning the refrigerator and reducing the number of times the refrigerator door is opened and shut.

Not leaving pilot lights constantly on, such as for the water heater and bath.

Choosing energy-saving products when buying replacement consumer electronics.

Using your car as less as possible for shopping, leisure, etc., and using public transportation, such as trains and buses, instead.

When driving, making sure to turn off the engine whenever possible.

Reduce (waste)

Not buying disposable products.

Not using plastic bags (taking your own shopping bag) and requesting that shops use simple wrapping.

Not buying items that will quickly go out of fashion or that you will soon be disinterested in.

As much as possible, not buying products that you do not need and instead renting or leasing products.

Sharing things you do not need with friends, acquaintances, etc.

Making compost out of food waste.

Using refillable products.

Choosing products that do not break easily and that will last a long time.

Repairing broken items to extend their lifespan.

Not throwing food away; for example, not leaving food half-eaten or buying or cooking too much.

Choosing stores that use simple wrapping and not using disposable food utensils (disposable chopsticks, etc.).

Not using food and drink utensils and containers that are used only once and are then thrown away.

Carrying your own chopsticks.

Reuse and recycle

Separating household waste into types and disposing of each type at the designated place.

(*Continued*)

Table 11.1 (Continued)

Washing bottles and similar items so they can be easily recycled.

Making full use of secondhand goods from recycling shops, bazaars and flea markets, Internet auctions, etc.

Buying products that use reusable containers, such as refillable bottles for beer and milk.

Using items you no longer need for other purposes, such as making rags out of old clothes.

Buying products made from recycled materials.

Cooperating with shops that collect recyclables, such as trays and milk cartons.

Cooperating with shops that collect small electronic equipment, such as mobile phones.

Ecosystems

Selecting and buying seasonal and local produce.

Actively interacting with nature by going outdoors and observing the living things around you.

Participating in eco-tours (guided nature experiences).

Participating in nature conservation and beautification activities.

Taking responsibility to nurture living things.

Prioritizing buying eco-friendly products.

Talking with your family and friends about nature and living things.

Utilizing methods that help mitigate global warming, such as saving electricity and setting an appropriate temperature for your air conditioner.

Table 11.2 Examples of pro-environmental behaviors (translated from Aoki et al. [2012])

1. Separating and discarding garbage according to the rules of your local area.
2. Recycling milk cartons, trays, etc.
3. Reusing disused items, such as giving them to bazaars, acquaintances, etc.
4. Carrying around your own water bottle and cup and limiting your use of disposable containers.
5. Carrying around your own chopsticks and reducing your use of disposable chopsticks.
6. Choosing and buying products with an eco-mark and made from recycled products.
7. When shopping, examining products closely and not buying any items that you do not need.
8. Using an item for a long time by taking care of it and by repairing when it is broken.
9. Choosing, buying, and consuming local produce.
10. Choosing and buying organic produce or organically grown products.

11. Refusing plastic bags and excessive wrapping.
12. Refusing disposable chopsticks, spoons, etc.
13. Using refillable products.
14. Spending time with your family in the same room (to save electricity).
15. Reducing the use of standby power by turning off appliances at the main power source when not in use.
16. Diligently cleaning the filters of air conditioners and vacuum cleaners.
17. Setting temperatures so that the air conditioner is not too cold and the heater is not too warm.
18. Using the bath continuously and not reheating it.
19. Adjusting the temperature of the heated toilet seat and keeping the toilet lid down.
20. Using curtains to block direct sunlight as well as to keep heat from escaping.
21. Carrying out green activities, such as planting trees, preparing flower beds, and installing green curtains.
22. Doing all your laundry at once to reduce the number of times you use the washing machine.
23. Diligently turning the water off when it is not needed while you are washing or showering.
24. Not leaving any food behind on your plate.
25. Diligently turning off the power supply to your TV and PC.
26. Diligently turning off room lights.
27. Pulling the plug out of the socket when you are not using the electric kettle for a long period of time.
28. Disposing of food waste at home, such as by composting or burying in the garden.
29. Making sure the stove flame is not misaligned with the pan, kettle, etc.
30. Squeezing out and throwing away as much of the water content of food waste as possible.
31. When cooking, trying to use up all of the ingredients and to reduce the amount of gas and electricity you use.
32. Not crowding the refrigerator and reducing the number of times you open and close its door.
33. Adjusting the temperature of the refrigerator according to the temperature of the room.
34. Not printing out things you do not need and reducing the frequency of printing.
35. Using both sides of the paper when printing.
36. Limiting your use of elevators, such as by using stairs.
37. Using the other side of paper for memos and printer paper.
38. Utilizing car-sharing and ride-sharing.
39. Using bicycles and public transport, such as trains, instead of cars.
40. Using carbon-offset products, green-power-certified products, eco-funding products, etc.
41. Donating to environmental conservation activities, etc.
42. Buying new and/or replacingsolar panels.
43. Buying new and/or replacing a solar heating system.
44. Buying new and/or replacing fuel cell equipment.

(Continued)

Table 11.2 (Continued)

---

45. Buying new and/or replacing wind power generation equipment.
46. Buying new and/or replacing a rainwater utilizing device.
47. Buying new and/or replacing equipment for composting food waste.
48. Buying new and/or replacing equipment to improve thermal insulation (insulating film, etc.)
49. Buying new and/or replacing double-glazed windows.
50. Buying new and/or replacing bulb-type fluorescent lights and LED lights.
51. Buying new and/or replacing an energy-efficient TV.
52. Buying new and/or replacing an energy-efficient boiler.
53. Buying new and/or replacing an energy-efficient refrigerator.
54. Buying new and/or replacing an eco-car (a hybrid car, etc.).
55. Ensuring that tires have the correct air pressure.
56. Not driving an unnecessarily loaded car.
57. Not idling, revving, or suddenly accelerating your car, and driving at a constant speed.

---

reducing the opening and closing of refrigerator doors. Thirteen behaviors focus on reducing the use of products, such as diligently turning off lights, using a product together with more than one person, and consuming reusable products. Other behaviors include properly maintaining products, separating recyclables, using public transportation, and donating used goods.

However, not all of these pro-environmental behaviors can be subsidized. This is because of the inherent features of the behaviors, not limitations in the subsidy. For instance, diligently turning off the TV when it is not being watched represents an at-home behavior that cannot be observed by others. The same can be said of reducing the opening and closing of a refrigerator door and using a product for an extended period. In contrast, the purchase of an energy-efficient product is a behavior that can be observed and confirmed by a retailer. Moreover, information on the behavior can be easily registered and stored at sales points through the retailer's cashier system.

Behaviors associated with product transactions require the involvement of a receiver or provider, which renders the behaviors more identifiable and applicable to a pro-environmental consumer subsidy program. An example of such a behavior other than product purchase is cooperation in the sorting of recyclables. Adachi Ward in the Tokyo metropolis in Japan collects used PET (polyethylene terephthalate) bottles and rewards points to citizens who cooperate with the sorted collection of PET bottles. However, some sorted collection systems are not suitable for a consumer subsidy program because participation in a curbside collection system of recyclables cannot be monitored unless citizens write their name on the bags of recyclables that they sort. The ability to distinguish the consumer, to some extent, influences the applicability of a consumer subsidy program.

Table 11.3 Number of behaviors within the scope of the Eco-Action Point program

Product purchase	Service utilization	Other behaviors	Total
82	59	49	190
(43%)	(31%)	(26%)	(100%)

The rebate program for energy-efficient appliances (the eco-point program for appliances) discussed in Part II, the eco-car subsidy program discussed in Part III, and the eco-point program for housing addressed in Chapter 10 all target product purchases. Behaviors observable by others, and for which the consumer can be distinguished, have a greater level of applicability to consumer subsidy programs. In Japan, another subsidy program, the Eco-Action Point (EAP) program, rewards points for various pro-environmental behaviors and eco-product purchases (see Box 11.1 for a detailed explanation of the EAP program). Table 11.3 summarizes the pro-environmental behaviors targeted by the EAP program. Among the 190 behaviors, 43 percent are product purchase oriented and 31 percent are product use oriented. The purchase of products/services is the most common program behavior. Other behaviors include participating in events and activities; visiting a store using an eco-friendly mode of transportation, such as walking; refusing unnecessary free services at hotels and shops; cooperating in sorted collection of recyclables; recording and reporting of pro-environmental housekeeping; joining pro-environmental organizations; acquiring a qualification relating to environmental activities; and donating to environmental causes. Many of these behaviors cannot reduce the environmental load directly, but do have some positive effects in the long-term or indirect effects that reduce environmental loads.

Box 11.1 Outline of the Eco-Action Point program

The Eco-Action Point (EAP) program is composed of four parties: members, fund providers, platforms, and providers of exchange products. In principle, the last party shall also provide funds as well as fund providers.

- Members receive eco-action points when they carry out eco-actions: purchase eco-products, use eco-services, and perform eco-activities. By registering the points, they can collect eco-action points, which they can then redeem for products and services.
- Fund providers apply to register on platforms eco-products and/or eco-services that are eligible for the EAP program. If their registration is approved and they provide funds for the eco-action points, then they

can sell and supply the eco-products and eco-services with eco-action points.

- The platform manages the eco-action points and the funds, registers eligible eco-actions, and provides information on members' balance of eco-action points as well as information about the reduction of environmental loads.
- Providers of exchange products register exchange products and/or services that can be redeemed for eco-action points on the platform.
- The Ministry of the Environment (MOE) approves the use of the name of EAP and its label for the main platform, and also monitors and evaluates the performance of the whole program.

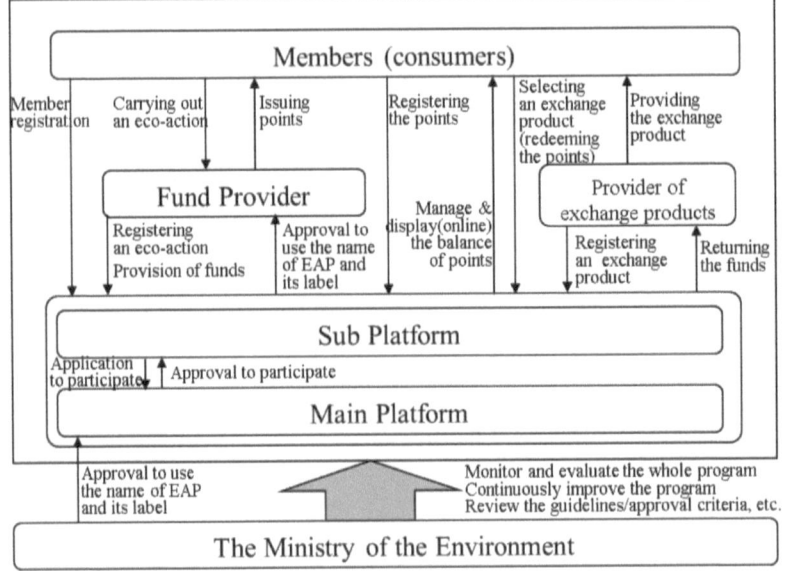

Figure 11.B1 The basic scheme of the Eco-Action Point (EAP) program

Source: The Eco-Action Point program guidelines issued by MOE (2012), translated by the author.

Table 11.4 summarizes the pro-environmental behaviors that fall within the scope of subsidy programs and that are included in the EAP program list. A behavior is an identifiable transaction where a product or money is given or received; behaviors that involve the transfer of information are more difficult to identify.

Pro-environmental behaviors that fall within the scope of subsidy programs tend to include a product, money, and/or information transfer between a

Table 11.4 Types of pro-environmental behaviors within the scope of subsidy programs and their features

Behavior	Item to be transacted			Place	Consumer distinction	Remarks
	Product	Money	Information			
Eco-product purchase	Product purchased	Payment	–	–	Easily identifiable	Applicable for traditional and Internet purchases
Eco-service purchase	Depends on service provided	Payment	–	At the site of service	Easily identifiable	
Participation in eco-event a) Charged	Entrance ticket	Payment	–	At entrance gate	Easily identifiable	A requirement for the subsidy is the purchase of an entrance ticket or passing through an entrance gate.
b) Free	–	–	–	A designated site in the venue	Identifiable	A requirement for the subsidy is to visit a designated site. Repeat visits should be identifiable to eliminate subsidy exploitation.
Visiting stores via eco-transportation	–	–	–	A designated site in the store	Identifiable	
Refusal of free services	Depends on service provided	–	–	At the site of service	Identifiable	An example of free services is the provision of free plastic bags at a shop. Refusal of a free service can reduce the associated environmental load and reduce provider costs.
Cooperation in sorted collection	Recyclables	–	–	Designated collection sites	Difficult in some cases	Subsidy programs cannot be applied to some collection systems because it is difficult to identify participating individuals.
Reporting	–	–	Declared pro-environmental behaviors	–	Difficult	It is difficult to verify behaviors. A subsidy program has to rely on self-declaration.
Qualification acquisition	For example, certification	Payment	–	–	Identifiable	Acquisition cannot reduce environmental loads. This addresses subsequent long-term activities and their outcomes.
Donation	–	Donation	–	–	Identifiable	Donating people/organizations are not necessarily conducting pro-environmental behaviors. This has a social rather than an environmental impact.

consumer and another individual. Alternatively, the behaviors may reflect consumer behavior at a certain location. In some cases, a behavior cannot be identified. Subsidy programs then must prevent the exploitation of subsidies by dishonest consumers.

How can a subsidy program operator subsidize behaviors that are not identifiable? A program could rely on self-declaration; however, fraud would subsequently reduce the size of the reduction effect on environmental loads. Such an approach is not sustainable and, if applied, should be on a short-term basis with the educational aim of providing information on pro-environmental behaviors.

11.3. Is it a pro-environmental behavior?

11.3.1. Assessment based on life cycle thinking

The purchase of eco-products is one of pro-environmental behaviors that can be supported by a consumer subsidy program. The environmental load caused by a product is typically assessed by life cycle assessment (LCA) or similar methods based on a life cycle approach. LCA is a methodology that assesses environmental loads generated from all of the life stages of a product and service, including those stages involved in the manufacturing of raw materials for the product, product use, and waste management when the product is discarded. According to the ISO definition (ISO 14040, 2006), LCA represents a "compilation and evaluation of the inputs, outputs and the potential environmental impacts of a product system throughout its life cycle." LCA has been used for identifying hot spots – the product life stages and processes having the greatest environmental loads – and for comparing the environmental performance of products and services.

For consumable goods, the purchase of products with greater environmental performance is considered a pro-environmental behavior. A subsidy program, therefore, targets the difference in environmental performance between a conventional, average product and a new environmentally-friendly product. However, for consumer durables, the environmental performance of a new product purchased as well as the performance of the product in possession to be replaced must be considered to determine whether the replacement represents a pro-environmental behavior. This is because consumer durables can be used for an extended period without replacement. That is, it is critical to compare the environmental load of the product in possession and the load created from the use of the new product. Such a comparison is, however, not simple, and we have to consider various factors concerning product replacement. For instance, it may be preferable not to replace a product in terms of the extent of environmental load if the product in possession exhibits high-energy performance at the time of purchase. Even if environmental loads from the use of a product are generated, greater environmental loads may be generated at the production stage of a new replacement product. Thus, when replacing a product, the environmental loads

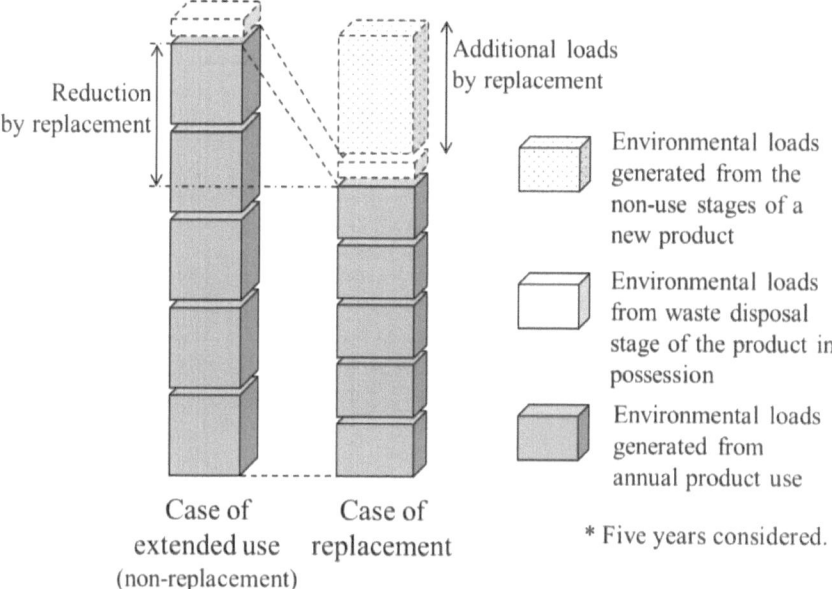

Figure 11.1 Environmental loads generated in the cases of product replacement and non-replacement (extended use)

generated from the production of the new product, the disposal (including recycling and intermediate/final treatment) of the product in possession, and the use of the product in possession must be considered.

Figure 11.1 illustrates these environmental loads. Two cases are compared. One is a case in which a product in possession will be used for five additional years, and the other is a case where the product is replaced with a new, energy-efficient product. The lower five grey boxes represent the amount of environmental load generated from product use for an additional five years (the left represents an existing product in possession, and the right represents the new product). The difference is the environmental gain obtained by the existing product's replacement with an energy-efficient product. The dotted small white boxes represent the amount of environmental load generated from the waste disposal stage of the product in possession. The large dotted box at the top of the right column represents the allocated amount of the environmental load from the non-use stages (production and waste disposal) of the new product. The allocation was made for five years using the average product life span (i.e. the amount of environmental load divided by average product life span and then multiplied by five).

11.3.2. Product category

Table 11.5 categorizes consumer durables into three types of environmental loads. The products in category 1 use a substantial percentage of the total life

Table 11.5 Categories of consumer durables (retrieved and modified from the Ministry of Environment [2010])

Product Category	Category 1	Category 2	Category 3
Feature	Consumption at the product use stage accounts for a large percentage of the product's total energy consumption.	Energy consumption at the product use stage is greater than that of the production and waste disposal stages.	No energy is consumed at the product use stage.[a]
Example	TVs, air conditioners, washing machines, refrigerators, and cars.	Digital cameras, mobile phones, game consoles, and personal computers.	Furniture, clothes, books, and bicycles.
Product replacement decision	Case specific.	Extended use is preferred.	Extended use is preferred.

a Energy is consumed for maintenance, which is not considered here. For instance, replacement parts consume energy (i.e. energy is required to produce the parts).

cycle energy consumption of the product during the use stage. The energy efficiency of product use has improved in the last decade; therefore, replacement of these products with a new, energy-efficient product reduces energy consumption and other associated environmental loads during the product use stage. However, product replacement increases energy consumption and other environmental loads, such as resource consumption generated from the production stage of the new product. A detailed assessment is required to determine whether the replacement of products in category 1 is preferable from an environmental perspective. Subsidy programs for the purchase and replacement of these products should be designed based on the insights of such an assessment. The products in category 2 use a substantial percentage of the total life cycle energy consumption during the production and waste disposal stages. Product replacement leads to increases in energy and resource consumption associated with product production even if energy consumption at the product use stage is improved. Thus, the extended use of products is preferred. Products in category 3 are products that do not consume energy at the product use stage. These products should also be used for an extended period from a basic environmental perspective.

Figure 11.2 shows the composition of Japanese household electricity consumption derived from hourly data. The four items – refrigerator, TV, lighting, and air conditioning – account for approximately 45 percent of electricity consumption. These items belong to product category 1. Electricity consumption of these items should be reduced, but the replacement of these products should be determined on a case-by-case basis.

Figures 11.3 and 11.4 illustrate the percentage of greenhouse gas emissions from the use stage of electrical and electronic equipment (the sum of "Electricity"

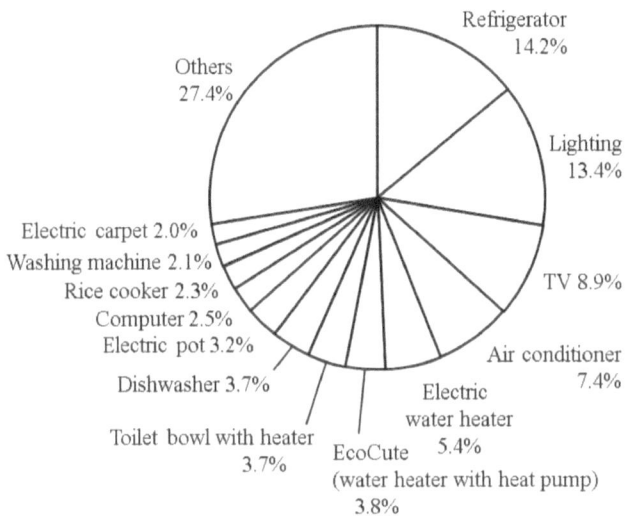

Figure 11.2 Electricity consumption of Japanese households (retrieved and translated from the Advisory Committee for Energy [2011])

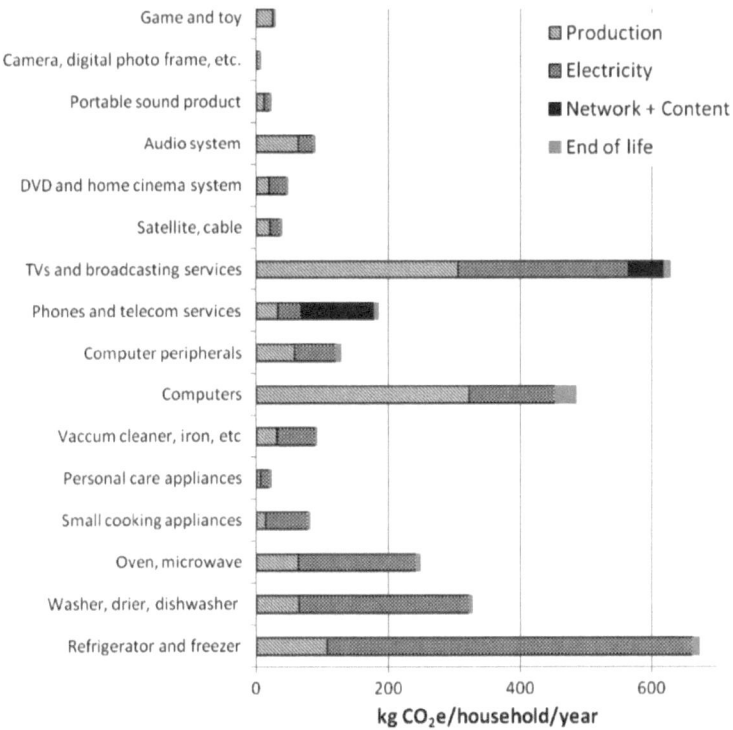

Figure 11.3 Greenhouse gas emissions from the consumption of electrical and electronic equipment by Norwegian households (Hertwich and Roux 2011). Reprinted with permission from Environmental Science & Technology. Copyright © 2011. American Chemical Society.

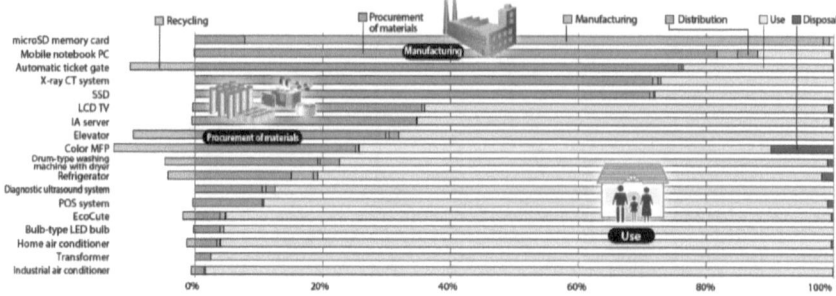

Figure 11.4 Percentage of CO$_2$ emissions from different stages of the life cycle of Toshiba Group's products (Toshiba 2013)

Note: Recycling can reduce CO$_2$ emissions, and the percentage of recycling having negative values is shown to the left of "0%." SD: secure digital, PC: personal computer, CT: computed tomography, SSD: solid-state drive, LCD: liquid crystal display, IA: Intel architecture, MFP: multifunction printer, POS: point of sale, LED: light-emitting diode.

and "Network + Content" in Figure 11.3 correspond to the product use stage. In Figure 11.4, the least-shaded bars represent the use stage). Greenhouse gas emissions can differ even if the same amount of electricity is consumed when the composition of power sources in a country differs. Figure 11.3 represents Norway, and Figure 11.4 represents Japan. However, there are similarities. Washing machines, dishwashers, and cooking appliances generate greenhouse gases largely during the product use stage, whereas computers generate greenhouse gases mainly during the production stage.

Payback time is an idea relating to replacement of products in categories 1 and 2. Payback time was originally used in the finance field to refer to the period required to recover the cost of an investment. The payback time, p, is expressed in Eq. 11.1 as the cost of investment, I, divided by the annual profit gained from the invested project or product R^*. Until the payback period has elapsed, the net profit of the investment is not realized.

$$p = I/R^* \tag{11.1}$$

Here, I represents investment and R^* represents annual recovery.

This idea of payback time is used for environmental load. For example, with respect to solar panel installation, I represents the energy consumption of the production and installation of the solar panels, R^* represents the annual energy produced by the solar panels, and p is an indicator showing whether a power generator produces net energy. Before the payback time has elapsed, net energy production (for the power-generating company, net profit) is not realized and the power generator should not be replaced.

Ordinary products do not produce energy at the use stage but rather consume energy. Consequently, the expression "payback" is not appropriate and should be avoided. The same form of the equation, Eq. 11.2, provides another period, T, with a different connotation. Period T represents a time when an environmental load at the use stage equals that of the non-use stages. That is, T distinguishes product category 1 and category 2 in Table 11.5. If the product life span is shorter than period T, an environmental load at the non-use stage is greater than the load at the product use stage, implying that even though environmental load at the product use stage is reduced, the reduction will not exceed the additional amount of environmental load at the non-use stages. Measures for energy consumption reduction should be addressed to the production stage rather than that the product use stage.

$$T = E_M / E_U^*$$ (11.2)

where E represents environmental load, * represents the annual value, and the subscripts M and U represent the manufacturing stage and use stage, respectively.

Table 11.6 compiles the T periods of several products used in Japan. The T periods with respect to energy consumption are approximately one year or less and, in terms of CO_2 emissions, less than three years. Laptop PCs (personal computers) are exceptional, and period T in this case is approximately five years. Therefore, replacing electrical home appliances less than every three years would not be environmentally preferable, and replacing laptop PCs less than every five years should be avoided with respect to the reduction of CO_2 emissions.

11.3.3   *The product replacement decision for category 1*

When deciding whether to replace products in category 1, the following factors should be considered: (1) the difference in generated environmental loads at the

Table 11.6 Periods when an environmental load from non-use stages is equal to that of the use stage, T, and the average lifespan of consumer durables used in Japan

	Period T (year)		Average Lifespan (years)
	Energy Consumption	CO₂ Emission	
Air conditioner[a]	0.4	0.6	15
Refrigerator[a]	1.0	2.8	12[a]
TV[a]	1.3	2.5	10[a]
PC (desktop without display)[b]	1.48	2.32	4.0[c]
PC (laptop)[b]	3.42	5.28	

a Calculated from data provided by Tasaki et al. (2013).
b Calculated from LCA data published by producers, which were surveyed by Hinodeya Institute for Eco-life (2010).
c Economic lifespan.

product use stage between a new product and the existing product and (2) the generated environmental loads at the non-use stages of a new product, as demonstrated in Figure 11.1. This decision is complex, and Tasaki et al. (2013) examined it in detail. Environmental loads for the standard (non-replacement) case (E_S) and the replacement case (E_R) in Figure 11.1 are expressed by Eqs. 11.3 and 11.4, respectively.

$$E_S = E_U^{O*} \cdot y_{av} + E_W^O \tag{11.3}$$

$$E_R = E_U^{O*} \cdot y_R + E_W^O + E_M^N \cdot \frac{(y_{av} - y_R)}{y_{av}} + E_U^{N*} \cdot (y_{av} - y_R)$$
$$+ E_W^N \cdot \frac{(y_{av} - y_R)}{y_{av}} \tag{11.4}$$

where E represents environmental loads; y_{av} represents the average lifespan of the product in years; y_R represents the duration of product possession at the time of the replacement decision; superscripts O and N represent a currently owned product and a new energy-efficient product, respectively; * represents the value of annual environmental loads; and subscripts M, U, and W represent the manufacturing, use, and waste disposal stages, respectively.

The condition that the replacement of a product is preferable is expressed as

$$E_S - E_R > 0 \tag{11.5}$$

Substitute this inequality with Eqs. 11.3 and 11.4:

$$E_U^{O*} \cdot (y_{av} - y_R) - E_M^N \cdot \frac{(y_{av} - y_R)}{y_{av}} - E_U^{N*} \cdot (y_{av} - y_R) - E_W^N \cdot \frac{(y_{av} - y_R)}{y_{av}} > 0 \tag{11.6}$$

Remove $(y_{av} - y_R)$ and divide the sides by E_U^{O*}:

$$1 - \frac{E_M^N}{E_U^{O*}} \cdot \frac{1}{y_{av}} - \frac{E_U^{N*}}{E_U^{O*}} - \frac{E_W^N}{E_U^{O*}} \cdot \frac{1}{y_{av}} > 0 \text{, and then}$$

$$\frac{E_U^{N*}}{E_U^{O*}} < 1 - \frac{E_M^N + E_W^N}{E_U^{O*} \cdot y_{av}} \tag{11.7}$$

Multiply both the denominator and numerator of the second term on the right side by $E_M^O + E_W^O$:

$$\frac{E_U^{N*}}{E_U^{O*}} < 1 - \frac{E_M^N + E_W^N}{E_M^O + E_W^O} \cdot \frac{E_M^O + E_W^O}{E_U^{O*} \cdot y_{av}} \tag{11.8}$$

Then, ε, Υ, and ϕ are defined as $\varepsilon = E_U^{N*} / E_U^{O*}$, $\gamma = \frac{E_M^O + E_W^O}{E_U^{O*} \cdot y_{av}}$, and $\phi = \frac{E_M^N + E_W^N}{E_M^O + E_W^O}$. ε represents the improvement rate of environmental loads at the product use stage

(–), γ represents the ratio of environmental loads at the non-use stages to that of the use stage (–), and ϕ represents the improvement rate of environmental loads at the non-use stages (–). For cases where product replacement is preferable, environmental loads are expressed by Eq. 11.9.

$$\varepsilon < 1 - \phi \cdot \gamma \qquad (11.9)$$

According to Eq. 11.9, replacement becomes preferable as the contribution of environmental loads at the product use stage becomes greater (γ is smaller), as the improvement in environmental loads at the use stage becomes greater (ε is smaller), and as the improvement in environmental loads at the non-use stages increases (ϕ is smaller). Consequently, replacement becomes preferable. Figure 11.5 shows the lines of Eq. 11.9 and the area where replacement is preferable. Hereinafter, the line is referred to as the IEL (iso-environmental-load) line. In cases where $\phi = 1$, the x-intercept of the IEL line is 1.0, which implies that there is no area where replacement is preferable when $\gamma \geq 1.0$. This corresponds to the insight that products in category 2 should not be replaced until γ becomes 1.0 at its shortest length. When $\gamma < 1.0$, both ε and γ determine whether replacement is preferable and, therefore, we must determine whether replacement is preferable on a case-by-case basis. If $\varepsilon + \gamma$ is smaller than 1.0, replacement is preferable. Because γ is larger (i.e. the contribution of environmental loads at non-use stages is greater), a smaller ε (i.e. greater reduction of environmental loads at the product use stage) is necessary for eco-friendly replacement of products. In Figure 11.5, the x-intercept depends on ϕ, which we have not discussed previously. If the environmental loads generated from the non-use stages doubled ($\phi = 2.0$), the area where product replacement is preferable – that is, the number of conditions where replacement is preferable – is halved.

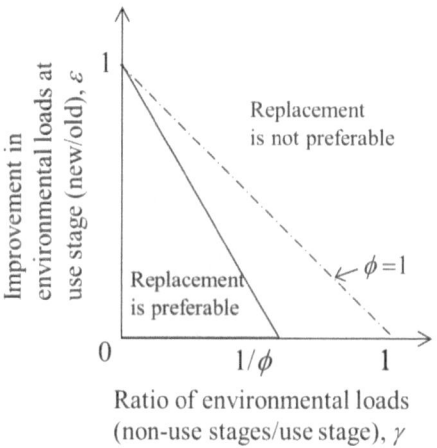

Figure 11.5 The iso-environmental-load line (Eq. 11.9) for the replacement of a product (Tasaki et al. 2013)

Parameter ε takes various values for energy consumption. In many cases, ε is less than 1.0 because energy efficiency has been improving in recent years. However, a new additional function may consume more energy. That is, consumers may purchase a larger product that consumes more energy. Flat-panel TVs are currently popular, and a larger flat-panel TV can replace a CRT TV in the corner of a room (e.g. a 21-inch CRT TV can be replaced by a 32-inch flat-panel TV). A 42- to 46-inch flat-panel TV can replace a 29-inch CRT TV in the equivalent space. Figure 11.6 shows the values of ε when a 25- to 29-inch TV replaced a new flat-panel TV in 2008 and in 2013. Figure 11.6a shows that ε is approximately 1.0 when a product purchased 10 years ago is replaced with a 15-inch-larger flat-panel TV (Point A) and is approximately 1.2 when a product purchased 5 years ago is replaced (Point B). Figure 11.5 shows that there is no area where replacement is preferable when $\varepsilon \geq 1.0$. Therefore, replacing a TV with a 15-inch-larger TV was not preferable in 2008. However, this was not the case in 2013. Figure 11.6b shows that ε is approximately 0.8 when a product purchased five years ago is replaced and the size of a new product is less than 40 to 44 inches (Point C). Replacing this product with a 15-inch-larger TV is now preferable because of improved energy efficiency.

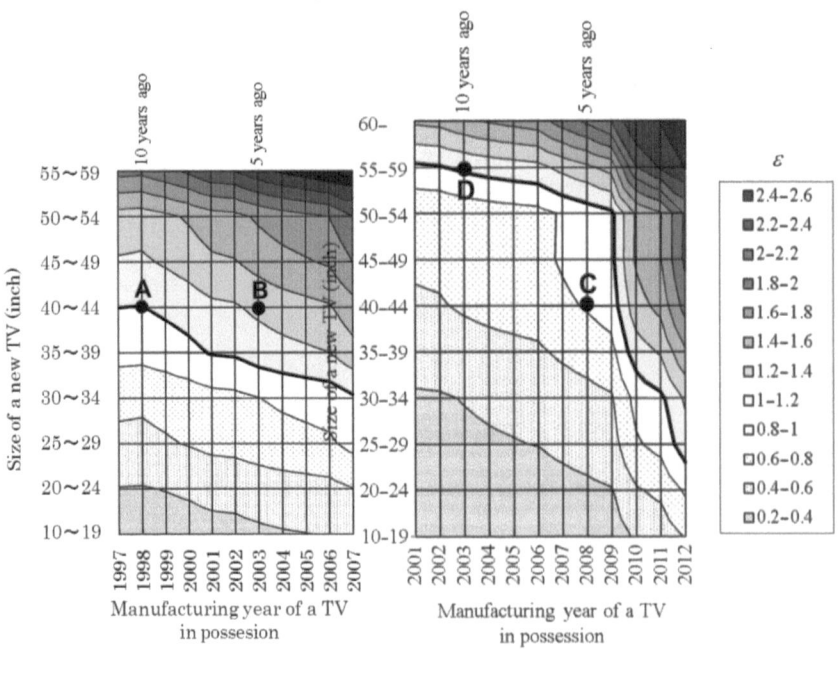

(a) Replaced in 2008   (b) Replaced in 2013

Figure 11.6. The ε values for replacement of a 25- to 29-inch TV (CRT: 2001–7, LCD: 2008–12) with a new-flat panel TV of different sizes in 2013

Note: The bold lines show $\varepsilon = 1.0$.

Source: Graph (a) is from Tasaki et al. (2013).

Another parameter to consider is γ, which represents the inverse of the contribution of energy consumption at the product use stage. This value, therefore, differs according to consumer use of the same category of a product. Table 11.7 summarizes the data that are related to the duration of daily/annual use of TVs and air conditioners in Japan. For instance, consumers in Japan watch TV for 4.5 hours a day on average, whereas some consumers watch TV for 1.4 hours and others for 15.8 hours. Table 11.8 shows that the value of ε can be obtained from these data and γ by using Eq. 11.9. The obtained ε represents the value required for eco-friendly replacement of products. We can see that for ordinal (average) consumers, an ε value of 0.75 (i.e. a reduction rate of environmental loads is 25 percent) is required for the eco-friendly replacement of a TV. Figure 11.6b also shows that, approximately, a value of ε below the line at the bottom of the white area meets the condition. A 25- to 29-inch TV manufactured five years ago should not be replaced with a TV that is greater than 45 inches in size (Point C). A 25- to 29-inch TV manufactured 10 years ago should not be replaced with a TV greater than 55 inches in size (Point D). Consumers who do not habitually use the product (see -2σ values in Table 11.7) should replace a TV when ε is 0.21 (i.e. the reduction rate of environmental loads is 79 percent or 67 percent). The

Table 11.7 Daily/annual use of products in Japan (from Tasaki et al. [2013])

		TV (hours/day)	Air conditioner* (hours/year)
Use occasionally	−2σ value	1.4	252
↑	σ value	2.6	555
	Average	4.5	1,221
↓	+σ value	8.7	2,689
Use often	+2σ value	15.8	5,926

* The sum of cooling time and warming time.

Table 11.8 Improvement ratio of environmental loads from the product use stage required for the eco-friendly replacement of products

		TV		Air conditioner		Refrigerator	
		Energy consump-tion	CO_2 emission	Energy consump-tion	CO_2 emission	Energy consump-tion	CO_2 emission
Use occasionally	−2σ value	0.59	0.21	0.47	0.33		
	σ value	0.78	0.57	0.76	0.69		
	Average	0.87	0.75	0.89	0.86	092	077
	+σ value	0.93	0.87	0.95	0.94		
Use often	+2σ value	0.96	0.93	0.98	0.97		

25–29 inches on the vertical axis of Figure 11.6b shows that ε is larger than 0.2, and replacing a TV with the same size is not preferable in terms of environmental loads even if the TV was used for 10 years. Table 11.8 also shows information concerning the replacement of air conditioners and refrigerators. Refrigerators are in use 24 hours a day, regardless of the consumer.

11.3.4. Providing information to consumers concerning environmental performance in Japan

How can individual consumers obtain information for decisions concerning the eco-friendly replacement of products? How should the government and business sectors provide such information?

A unified label is used in Japan for product performance of electrical home appliances, including environmental performance. This labeling scheme is based on Article 86 of the Rational Use of Energy Act, which states:

> Business operators engaged in supplying energy to general consumers, business operators engaged in selling or renting buildings, business operators engaged in retailing energy-consuming machinery and equipment, and other business operators capable of cooperating, through their business activities, in general consumers' efforts towards the rational use of energy shall endeavor to provide information that contributes to general consumers' efforts towards the rational use of energy, such as by making notifications on the status of energy use by consumers, by giving indications of the performance required for buildings to prevent heat loss through exterior walls, windows, etc. of the buildings and to realize the efficient utilization of energy for air conditioning systems, etc. installed in the buildings, and by giving indications of the performance of machinery and equipment in light of energy consumption.
>
> (From Japanese Law Translation created by the Ministry of Justice, Japan)

The labeling scheme was initiated in October 2006. Figure 11.7 shows the label. Retailers selling designated items are expected to display the label.

Since 2008, the MOE has provided detailed information concerning product replacement on the Internet. The MOE website is called "shin-kyu-san" (Figure 11.8). Consumers select a product in possession and a product to be purchased and then respond to questions concerning the use of the products on the website. The website then generates and displays bar charts of electricity consumption, electricity cost, CO_2 emission, and CO_2 absorption before and after replacement. However, the system does not take environmental loads generated from non-use product life stages into account. For decisions concerning the environmentally-friendly replacement of TVs, air conditioners, and refrigerators, all that consumers are required to do is to calculate the reduction rates of electricity consumption and CO_2 emission with the "shin-kyu-san" system and then compare them with the ε values in Table 11.8. Consumers do not require Figure 11.6 when using the system.

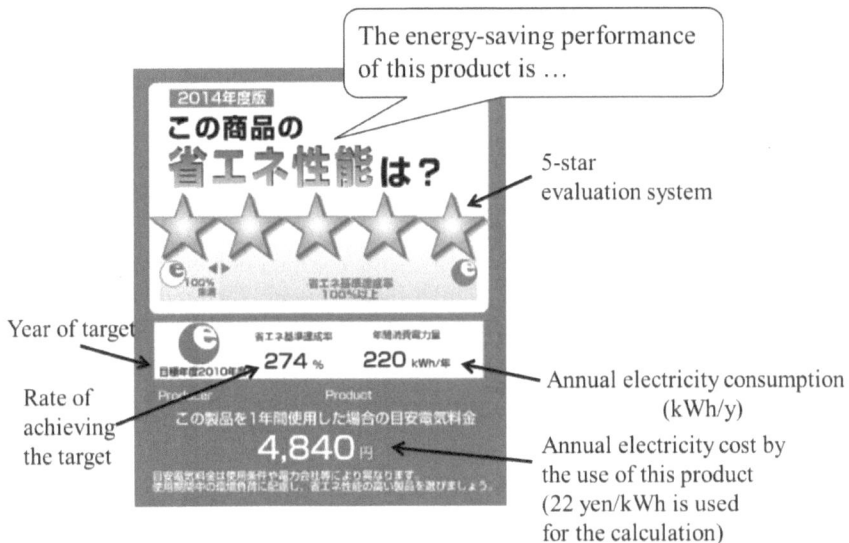

Figure 11.7 The unified label used in Japan for electrical home appliances

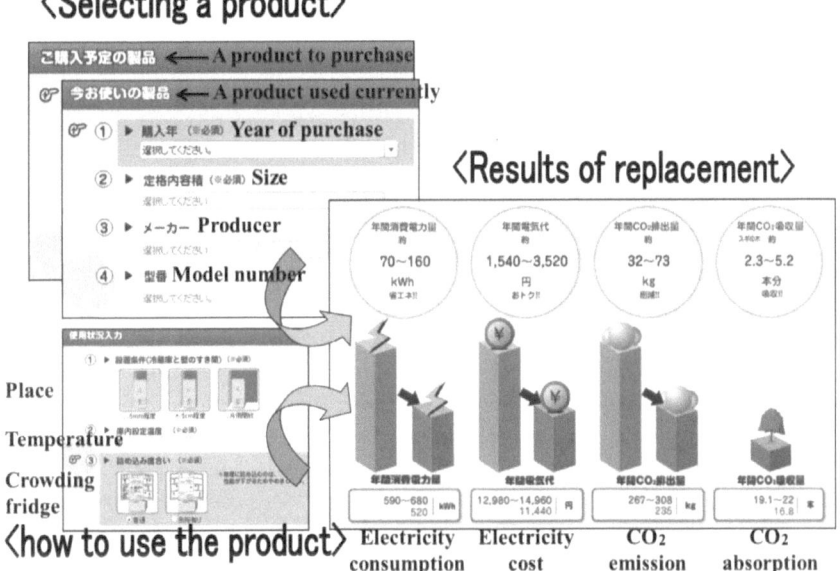

Figure 11.8 A website to support product replacement decisions with regard to electrical home appliances ("shin-kyu-san" provided by the Ministry of Environment)

11.3.5. Rebound effect

When life cycle thinking is applied to product replacement in an extended way, the rebound effect must be examined. The rebound effect has been discussed with respect to energy since the 1980s. As discussed earlier in this chapter, improvements in energy efficiency and decreases in the price of fuels may result in increased energy consumption. A number of definitions of the rebound effect have been offered. Madlener and Alcott (2009) defined the effect in the following way: "Rebound is the additional energy consumption enabled by energy efficiency increases." Several typologies distinguish direct rebound from indirect rebound. According to Madlener and Alcott (2009), direct rebound implies an increase in the demand for the same product or service, whereas indirect rebound implies an increase in the demand for a different product or service.

Considering economic factors, a comparison of environmental loads between product replacement and the extended use of a product requires a system boundary that includes the money saved from extended use. A consumer will spend the money saved, and this consumption generates additional environmental loads. Even if a consumer deposits the money into a bank, the saving behavior stimulates domestic fixed capital formation through the bank, and the capital formation generates environmental loads. Additional generation of these environmental loads is an indirect rebound effect associated with product replacement (or product non-replacement).

Kagawa et al. (2006) examined this type of the rebound effect for cars. The extended use of cars for an additional year reduced waste generation by 249,000 tons without considering rebound effects. When rebound effects are considered for two cases, namely (a) consumers who spend the surplus money on the purchase of other products and services and (b) consumers who save the money, then 49,000 tons and 443,000 tons of waste are generated as a rebound effect for (a) and (b), respectively. In case (b), the benefit from the reduction of waste is offset by the rebound effect, resulting in a net increase in waste generation.

Subsidies for pro-environmental behavior are likely to generate rebound effects. In this case, a subsidy enables consumers to spend more and generates additional environmental loads.

However, consideration of rebound effects is not necessarily required for product replacement, although rebound effects may offset reductions in environmental loads. The behaviors of product replacement and the spending of surplus money are independent. This chapter has explained that product replacement implies that consumers examine the values of the parameters of ε and γ and decide whether replacement is preferable. To be of environmental benefit, the spending of surplus money would require that environmental loads caused by various consumptive behaviors be estimated and publicized in advance and that consumers select behaviors with smaller environmental loads.

A points reward system rather than a consumer subsidy program may require that policy makers consider rebound effects because a points reward system encompasses both product replacement and associated behaviors – the use of points, which includes the purchase of products and services. The associated behaviors are likely to cause rebound effects and generate environmental loads.

11.4. Summary

This chapter discussed (1) whether it is easy to determine the occurrence of certain behaviors and (2) to identify whether a behavior conducted was pro-environmental to broaden our understanding of consumer subsidy programs. We confirmed that the purchase of products and services is one of the most identifiable pro-environmental behaviors and that gathering and storing information on such behaviors is relatively easy. We elucidated a framework for product replacement in terms of environmental loads and found that the main parameters are (1) differences in environmental loads at the product use stage between a new product and a product in possession and (2) environmental loads generated during the production stage of a new product. We categorized products into three types and provided general guidelines for the product replacement of each category. We presented a method for the product replacement decision and noted that product replacement decisions are complex. Guidelines for the replacement of TVs, air conditioners, and refrigerators were discussed. Finally, we addressed rebound effects. The rebound effect may offset reductions in environmental loads; the dissemination of information on the behaviors that generate less environmental load is required.

References

Advisory Committee for Energy. (2011) Reference document #1 of the 17th meeting of the Group for Energy Saving Standards, 26 December, p. 18 (In Japanese).

E. Aoki, K. Kurisu, and K. Hanaki. (2012) 'Analysis of factors contribute to citizen's environmental behavior in local government', *Journal of Japan Society of Civil Engineers, Division G: Environmental Systems and Engineering* 68(6), 2012, II_165–II_176 (In Japanese).

E. G. Hertwich and C. Roux. (2011) 'Greenhouse gas emissions from the consumption of electric and electronic equipment by Norwegian households', *Environmental Science & Technology* 45, 8190–6.

Hinodeya Institute for Eco-life. (2010) 'A report on collection and analysis of data on environmental loads and product use of electrical home appliances and cars'. (In Japanese).

ISO 14040. (2006) 'Environmental management – Life cycle assessment – Principles and framework'.

Japanese Cabinet Office. (2005) 'Public opinion survey on energy', Available: <http://survey.gov-online.go.jp/h17/h17-energy/index.html> (accessed 24 August 2014) (In Japanese).

Japanese Cabinet Office. (2012) 'Public opinion survey on environmental issues', Available: <http://survey.gov-online.go.jp/h26/h26-kankyou/index.html> (accessed 24 August 2014) (In Japanese).

S. Kagawa, T. Tasaki, and Y. Moriguchi. (2006) 'The Environmental and Economic Consequences of Product Lifetime Extension: Empirical Analysis for Automobile Use', *Ecological Economics* 58 (1), 108–18.

R. Madlener and B. Alcott. (2009) 'Energy rebound and economic growth: A review of the main issues and research needs', *Energy* 34, 370–6.

Ministry of the Environment, Japan (MOE). (2010) 'A report on investigation of current state of distribution and waste management of electrical and electronic equipment and promotion of reuse', Available: <https://www.env.go.jp/recycle/report/h22-07.pdf> (accessed 24 August 2014) (In Japanese).

Ministry of the Environment, Japan. (2012) 'Guidelines for eco-action point program', Main report, Available: <http://www.env.go.jp/policy/eco-point/guideline/guideline_full.pdf> (accessed 24 August 2014) (In Japanese).

Ministry of the Environment, Japan. (n.d.) 'Shin-kyu-san: a navigation for replacement of energy-saving products', Available: <http://shinkyusan.com> (accessed 24 August 2014) (In Japanese).

Ministry of Justice, Japan. (n.d.) 'Japanese law translation', Available: <http://www.japaneselawtranslation.go.jp/> (accessed 24 August 2014).

T. Tasaki, M. Motoshita, H. Uchida, and Y. Suzuki, (2013) 'Assessing the replacement of electrical home appliances for the environment – an aid to consumer decision making', *Journal of Industrial Ecology* 17 (2), 290–8.

Toshiba Corporation. (2013) 'Environmental Report 2013', p. 27, Available: <http://www.toshiba.co.jp/env/en/communication/report/pdf/env_report13_all.pdf> (accessed 21 August 2014).

Part V
Concluding part

12 Lessons for future environmental and industrial policies

Concluding remarks

Shigeru Matsumoto

12.1. Summary of chapters

To reduce household environmental impacts, an increasing number of environmental subsidy programs have been implemented in both developed and developing countries. Despite the recent popularity of these subsidy programs, their impacts on consumer behaviors and the environment have not been examined in detail. In this book, the authors have studied environmental subsidy programs recently introduced in Japan and have assessed their impacts.

Part I of this book summarized the points of debate on environmental subsidy programs. Economists explain that an externality can cause environmental problems. However, market failures such as asymmetric information and bounded rationality may also exist. The resolution of such market failures is often mentioned as a reason for why environmental subsidies are needed. In Chapter 2, after explaining the economic rationale for environmental subsidy programs, Matsumoto reported that subsidy programs increased the sales of energy-efficient products. A policymaker must allocate financial resources to implement subsidy programs. In Chapter 3, Sekiguchi conducted a historical review of Japan's environmental public finance and summarized the characteristics of recent environmental subsidy programs.

Part II of this book analyzed the effects of the eco-point program for appliances from several dimensions. In Chapter 4, Yamaguchi, Matsumoto, and Tasaki analyzed the effect of this program on consumer digital TV selection. They found that the program expanded the market share of energy-efficient digital TVs within each screen-size class but also increased the average screen size of digital TVs purchased. Further, durable home appliances have a product replacement cycle, and a certain proportion of old appliances are replaced with new ones even without a subsidy program. In Chapter 5, Morita and Arimura demonstrated that the estimated CO_2 reduction through this subsidy program becomes much lower than the government's estimate if the replacement cycle of refrigerators is incorporated. To assess the real impact of the subsidy program on CO_2 mitigation, it is necessary to compare household electricity consumption before and after the program. In Chapter 6, Mizobuchi and Takeuchi showed that after purchasing a new energy-efficient air conditioner households increased their usage of it during the summer. Nevertheless, they do find that total electricity consumption was reduced due to energy efficiency improvements.

In Chapter 7, Matsumoto, Morita, and Tasaki estimated hedonic price functions for air conditioners and examined how the subsidy program affected consumer valuation of the energy-saving function. They found that the program lowered consumer valuation of this function. Thus, the program aggravated the myopic behavior of consumers.

Part III of this book studied rebate programs for new-generation vehicles. In Chapter 8, Hirota and Kashima provided a comprehensive summary of vehicle registration and inspection policies in developing countries. By comparing those countries' policies with those of Japan, they highlighted the requirements for implementing programs to promote energy-efficient vehicles in developing countries. In Chapter 9, Iwata and Matsumoto analyzed data on Japan's used car market to evaluate the effectiveness of the rebate program for hybrid vehicles. They showed that despite owners of these vehicles driving much longer distances than owners of conventional gasoline vehicles, hybrid vehicles emitted less CO_2 due to their better fuel economy.

Part IV examined environmental subsidy programs from a broader perspective. The Japanese government subsidized the construction of energy-efficient "eco-friendly" houses. In Chapter 10, Iwata conducted an econometric analysis to evaluate the impact of the subsidy for eco-friendly houses. He showed that the subsidy program increased the number of eco-friendly houses, particularly in cold regions. In Chapter 11, Tasaki reassessed the eco-point program for appliances from a life cycle perspective. He argued that subsidy programs focused on the environmental burden at the use stage, and he showed that these burdens at the production and disposal stages could be substantial for some products. He then proposed a new energy efficiency criterion for future subsidy programs.

12.2. Lessons for future policies

The analyses in this book present several lessons for future environmental and industrial policies. All of the authors reported that the subsidy programs expanded the sales share of energy-efficient products. They also reported that CO_2 emissions from households have been reduced through the spread of environmentally-friendly products. Nonetheless, the analyses found that the cost-effectiveness of the environmental subsidy programs was highly questionable. The estimated costs to reduce 1 ton of CO_2 emission were 23,249 yen ($193.74) for hybrid vehicles and 27,200 yen ($226.67) for energy-efficient houses.[1] These costs are much higher than the CO_2 price reported in previous literature.

The cost-inefficiency in CO_2 reduction perhaps demands an additional policy objective for implementing an environmental subsidy program. In fact, we find that no program focuses only on environmental improvement and that all subsidy programs were introduced with multiple objectives, such as ensuring penetration of new technologies and recovery from recession.

This book further addressed some issues on system operation. Energy efficiency information is essential for a subsidy program for appliances. Vehicle registration information is necessary for a subsidy program to promote energy-efficient vehicles. Such information is not yet available in developing countries, and hence a system for collecting it would need to be established.

In many subsidy programs, products are classified into several categories according to their size. For instance, digital TVs are classified according to their screen size, refrigerators are classified according to their volume, and vehicles are classified according to their weight. A different energy efficiency criterion is used for each product category. In other words, relative efficiency criteria – rather than absolute efficiency criteria – are used. As we showed in the digital TV analysis in Chapter 4, the application of a relative efficiency criterion is sometimes problematic.

Present subsidy programs focus on the environmental burden at the use stage. Since subsidy programs accelerate the replacement of durable products, the environmental burdens at the production and disposal stages must be included in the efficiency assessment. An application of an alternative energy-efficient criterion based on life-cycle assessment is demanded.

Households benefit from environmental subsidy programs, and thus environmental subsidy programs continue to be used for greening household behavior. As we discussed in this book, however, the present subsidy programs have many deficiencies as environmental and industrial policies. The environmental subsidy programs must be further assessed and their deficiencies remedied.

Note

1 120 yen is converted into $1.

Index